W0235071

▶ **Policy-Oriented Technology Assessment Across Europe**

DOI: 10.1057/9781137561725.0001

Other Palgrave Pivot titles

Robert A. Stebbins: Leisure and the Motive to Volunteer: Theories of Serious, Casual, and Project-Based Leisure

Dietrich Orlow: Socialist Reformers and the Collapse of the German Democratic Republic

Gwendolyn Audrey Foster: Disruptive Feminisms: Raced, Gendered, and Classed Bodies in Film

Catherine A. Lugg: US Public Schools and the Politics of Queer Erasure

Olli Pyyhtinen: More-than-Human Sociology: A New Sociological Imagination

Jane Hemsley-Brown and Izhar Oplatka: Higher Education Consumer Choice

Arthur Asa Berger: Gizmos or: The Electronic Imperative: How Digital Devices have Transformed American Character and Culture

Antoine Vauchez: Democratizing Europe

Cassie Smith-Christmas: Family Language Policy: Maintaining an Endangered Language in the Home

Liam Magee: Interwoven Cities

Alan Bainbridge: On Becoming an Education Professional: A Psychosocial Exploration of Developing an Education Professional Practice

Bruce Moghtader: Foucault and Educational Ethics

Carol Rittner and John K. Roth: Teaching about Rape in War and Genocide

Robert H. Blank: Cognitive Enhancement: Social and Public Policy Issues

Cathy Hannabach: Blood Cultures: Medicine, Media, and Militarisms

Adam Bennett, G. Russell Kincaid, Peter Sanfey, and Max Watson: Economic and Policy Foundations for Growth in South East Europe: Remaking the Balkan Economy

Shaun May: Rethinking Practice as Research and the Cognitive Turn

Eoin Price: 'Public' and 'Private' Playhouses in Renaissance England: The Politics of Publication

David Elliott: Green Energy Futures: A Big Change for the Good

Susan Nance: Animal Modernity: Jumbo the Elephant and the Human Dilemma

DOI: 10.1057/9781137561725.0001

palgrave▶pivot

Policy-Oriented Technology Assessment Across Europe: Expanding Capacities

Edited by

Lars Klüver
The Danish Board of Technology Foundation, Denmark
Rasmus Øjvind Nielsen
The Danish Board of Technology Foundation, Denmark

and

Marie Louise Jørgensen
The Danish Board of Technology Foundation, Denmark

palgrave
macmillan

DOI: 10.1057/9781137561725.0001

First published 2016 by
PALGRAVE MACMILLAN

The authors have asserted their rights to be identified as the authors of this work in accordance with the Copyright, Designs and Patents Act 1988.

Palgrave Macmillan in the UK is an imprint of Macmillan Publishers Limited, registered in England, company number 785998, of Houndmills, Basingstoke, Hampshire, RG21 6XS.

Palgrave Macmillan in the US is a division of Nature America, Inc., One New York Plaza, Suite 4500 New York, NY 10004-1562.

Palgrave Macmillan is the global academic imprint of the above companies and has companies and representatives throughout the world.

Hardback ISBN: 978–1–137–56171–8
E-PUB ISBN: 978–1–137–56173–2
E-PDF ISBN: 978–1–137–56172–5
DOI: 10.1057/9781137561725

Distribution in the UK, Europe and the rest of the world is by Palgrave Macmillan®, a division of Macmillan Publishers Limited, registered in England, company number 785998, of Houndmills, Basingstoke, Hampshire RG21 6XS.

Library of Congress Cataloging-in-Publication Data is available from the Library of Congress

A catalog record for this book is available from the Library of Congress

A catalogue record for the book is available from the British Library

Contents

DOI: 10.1057/9781137561725.0001

List of Figures

▶

List of Tables

DOI: 10.1057/9781137561725.0003

Preface

This volume gives an updated picture of technology assessment (TA) in Europe and provides outlooks towards the establishment of a common European TA capacity for knowledge-based policy formation.

The volume gathers contributions from participants in the PACITA (Parliaments and Civil Society in Technology Assessment) project, which ran from 2011 to 2015. The volume is divided into three parts, which are preceded by Introduction that posits the expansion of TA capacities across Europe as a necessary supplement to existing European institutions, as well as a re-print of the so-called PACITA Manifesto, which urges policy makers at national and European levels to support such an expansion.

Part I of the book – Expanding Technology Assessment – pushes the concept of TA beyond its traditional limits and shows how TA may be institutionalized as a flexible system of collaborative efforts among a diverse range of actors across Europe. Chapter 1 examines existing TA institutions and their institutional roles and argues that TA as an umbrella term in fact applies to a broad range of 'TA-like' policy support functions that might meaningfully be termed 'policy-oriented technology assessment'. Chapter 2 gives an account of the practical efforts and successes within PACITA at expanding TA geographically to the participating countries, yet without formal parliamentary TA institutions, namely Belgium (Wallonia), Bulgaria, the Czech Republic, Hungary, Ireland, Lithuania and Portugal and shows the multiplicity of ways in which

TA may be adapted to the different national contexts. Chapter 3 takes the point of view of institutional entrepreneurs in the post-communist partner countries and depicts the process of adopting TA and adapting it to established national and organizational cultures. Chapter 4 conveys the outcomes of a series of parliamentary dialogues about the possible value of TA for parliaments in which parliamentarians called for a cross-European TA capacity to support reflexive European governance of science, technology and innovation. Chapter 5 sums up the decade-long experience in the TA field of developing cross-European forms of collaboration as well as the practical experiences gained in PACITA. Here, the common lesson is that not only is European collaboration in TA methodologically and organizationally feasible but it also adds greatly to the value of individual TA projects by allowing cross-national learning and comparison.

Part II – Exemplifying Cross-European Technology Assessment – digs into three example projects that were part of PACITA. Here, three different methodological approaches were applied on three highly policy-relevant topics. Chapter 6 describes a cross-national 'Future Panel' of parliamentarians from different member states, which facilitated a strategic learning process about the possible contributions of public health genomics to the healthcare systems of the future. Chapter 7 tells the story of a process of structured parallel national stakeholder dialogues about the future of ageing, which fed into and connected national and European policy debates on ageing society. Chapter 8 describes and analyses a cross-European citizen summit event concerning European policies on sustainable consumption, which made it clear that nationally rooted deliberation on highly complicated matters can serve as a filter to sort the real wishes of citizens out from the fears of decision makers about citizens' reactions to policy measures. Overall, the three chapters show that developing practices of cross-border TA is both practicable and valuable, but it needs to be institutionally rooted at national level – with European support.

The final part, Part III, of the book – Building Capacities for Cross-European TA – takes on the concrete question of how to proceed in establishing cross-European TA capacities. The chapters explain some of the steps already taken within the PACITA project and point to future perspectives for building on those efforts. These include the development of common training courses for practitioners (Chapter 9), educational seminars for policy makers and journalists (Chapter 10), international

DOI: 10.1057/9781137561725.0004

TA conferences (Chapter 11) and a common IT-platform for exchange and collaboration among practitioners (Chapter 12).

All contributions to this volume stand on the shoulders of the public deliverables of the PACITA project, which are available at www.pacitiaproject.eu.

DOI: 10.1057/9781137561725.0004

Acknowledgements

The PACITA (Parliaments and Civil Society in Technology Assessment) project was funded by the European Union's Seventh Framework Programme for research, technological development and demonstration under grant agreement number 266649, which made the project possible.

We wish to thank the network for European Parliamentary Technology Assessment (EPTA) for taking the initiative to launch PACITA. Special thanks also go to Member of the European Parliament, former Research Commissioner and former Chairman of STOA Philippe Busquin, as well as former Chairwoman of the Committee for Education, Research and Technology Assessment at the German Bundestag Ulla Burchardt, for driving debate on expanding technology assessment to all European member states, thereby providing a platform on which PACITA could unfold.

We are grateful to all those parliamentarians from the member states who involved themselves in the project and helped to further the debate. We are also thankful to the many stakeholders and citizens who chose to spend time participating in the example projects within PACITA.

Last but by no means least we wish to thank the international colleagues who supported the PACITA project with fresh insights and critical perspectives. For participation in the project's advisory panel, thanks go to Attila Havas, Gilbert Fayl, Nera Kuljanic, Theo Karapiperis, Timothy M. Persons and Wiebe Bijker. For taking time out to help shape this book, thanks go to Dionysia Lagiou, Hillary Sutcliffe, Josee van Eijndhoven and Karel-Herman Haegeman.

DOI: 10.1057/9781137561725.0005

The members of the PACITA consortium:

Fonden Teknologiraadet
The Danish Board of Technology
  Foundation Denmark

Institute of Technology Assessment,
  Karlsruher Institut für Technologie
Karslruhe Institute of Technology,
  Germany

Rathenau Instituut
The Rathenau Institute, Netherlands

Teknologiraadet
The Norwegian Board of Technology,
  Norway

Institut für Technikfolgen-Abschätzung,
  Austrian Academy of Sciences
The Institute of Technology Assessent,
  Austria

Applied Research and Communifactions
  Fund, Bulgaria

Instituto de Technologia Quimica e
  Biologica
Institute of Biology and Chemistry,
  Portugal

Institut Samenvlevning en Technologie
Institute for Society and Technology,
  Flandern, Belgium

Fundació Catalana per a la Recerca i la
  Innovació
Catalan Foundation for Research and
  Innovation, Catalonia, Spain

Zentrum für Technologiefolgen-
  Abschätzung
Swiss Centre for Technology Assessment,
  Switzerland

Ziniy Ekonomikos Forumas
Knowledge Economy Forum, Lithuania

Technologickê centrum AV CR
Technology Centre ASCR, Czech Republic

Univsertié de Liège, SPIRAL
University of Liège, SPIRAL Research
  Centre. Wallonia

Colàiste na hOllscoile Corcaigh
University College Cork, Ireland

Magyar Tudománuos Akadémia
Hungarian Academy of Sciences, Hungary

List of Contributors

Frédéric Adam, PhD, Associate Professor of Business Information Systems at University College Cork in Ireland and a visiting research fellow in the School of Economics and Management at Lund University, Sweden.

Mara Almeida, BSc, PhD in Developmental Biology, project manager at the Institute of Biological and Chemical Technology, Portugal.

Marianne Barland, Master in European Science and Technology studies, project manager at the Norwegian Board of Technology (NBT), Norway.

Danielle Bütschi, doctorate in Social Sciences and Economics, senior researcher and project manager at TA-Swiss, Switzerland.

Blagovesta Chonkova, MSc in Public Policy, project officer at ARC Fund, Bulgaria.

Zoya Damianova, MSc, MBA, program director responsible for strategic development at ARC Fund, Bulgaria.

Pierre Delvenne, PhD in Political Science, research fellow at SPIRAL Research Centre, University of Liège, Wallonia, Belgium.

Ciara Fitzgerald, PhD, is a lecturer in the Department of Accounting, Finance and Information Systems, senior postdoctoral research fellow, University College Cork, Ireland.

Katalin Fodor, International Relations Officer, the Hungarian Academy of Sciences, Hungary.

DOI: 10.1057/9781137561725.0006

Jurgen Ganzevles, MSc, PhD, senior researcher at the Rathenau Instituut, the Netherlands.

Julia Hahn, MA in Cultural Sciences, scientific staff member at the Institute for Technology Assessment and Systems Analysis (ITAS), Karlsruhe Institute of Technology, Germany.

Lenka Hebakova, MSc in Political Science, project manager at the Strategic Studies Department of the Technology Centre ASCR, the Czech Republic.

Leonhard Hennen, Dr. phil. In Sociology, senior researcher at the Institute for Technology Assessment and Systems Analysis (ITAS), Karlsruhe Institute of Technology, Germany.

Marie Louise Jørgensen, MSc in Public Administration, project manager at the Danish Board of Technology Foundation (DBT), Denmark.

Lars Klüver, MSc In Biology/Ecology, Director of the Danish Board of Technology Foundation (DBT), Denmark. Coordinator of the PACITA project.

Ventseslav Kozarev, MSc in Public Policy, MA in Political Science, researcher at ARC Fund, Bulgaria.

André Krom, PhD, senior researcher in the technology assessment department of the Rathenau Instituut, the Netherlands.

Edgaras Leichteris, Master's in Law and Management, PhD student in Public Administration, Director of Knowledge Economy Forum, Lithuania.

Katrine Lindegaard Juul, MSc in social anthropology, project manager at the Danish Board of Technology Foundation (DBT), Denmark.

Tomáš Michaelek, PhD in International Relations, research policy analyst in the Strategic Studies Department of Technology Centre ASCR, the Czech Republic.

Judit Mosoni-Fried, Dr oec. in Economics, senior researcher at the Institute for Research Organization, Hungarian Academy of Sciences, Hungary.

Michael Nentwich, Dr habil, Director of Institute of Technology Assessment, Austrian Academy of Sciences, Austria.

DOI: 10.1057/9781137561725.0006

Rasmus Øjvind Nielsen, MA, PhD student In Public Administration, researcher at the Danish Board of Technology Foundation (DBT), Denmark.

Linda Nierling, PhD in Sociology, researcher at the Institute for Technology Assessment and Systems Analysis (ITAS), Karlsruhe Institute of Technology, Germany.

Walter Peissl, PhD in Social Sciences, Deputy Director and a senior researcher at the Institute of Technology Assessment, Austrian Academy of Sciences, Austria.

Benedikt Rosskamp, MSc, PhD in Social and Political Science, researcher at SPIRAL Research Centre, University of Liège, Wallonia, Belgium.

Arnold Sauter, Dr rer. nat. in Zoology, vice-director of office of technology assessment at the German Parliament (TAB), Germany.

Constanze Scherz, Diploma in Social Sciences, scientific staff member at the Institute for Technology Assessment and Systems Analysis (ITAS), Karlsruhe Institute of Technology, Germany.

Stefanie B. Seitz, Diploma in Biology, PhD, scientific staff member at the Institute for Technology Assessment and Systems Analysis (ITAS), Karlsruhe Institute of Technology, Germany.

Dirk Stemerding, PhD, senior researcher at the Rathenau Instituut, the Netherlands.

Rinie Van Est, research coordinator and 'trendcatcher' at the Rathenau Instituut, the Netherlands.

DOI: 10.1057/9781137561725.0006

Introduction: On the Concept of Cross-European Technology Assessment

Rasmus Øjvind Nielsen and Lars Klüver

Abstract: *Nielsen and Klüver introduce the concept of cross-European technology assessment developed in the PACITA project, the layers of which are unfolded in the remaining chapters of this book. As a supplement to existing European institutions, cross-European technology assessment is a vision of a networked support system for national parliaments supplying process-support for knowledge-based and participatory policy making. As well as discussing the possible role of such a support system within existing European frameworks of policy formation, Nielsen and Klüver propose the necessity of capacity building modelled on the concept of cross-European technology assessment as a means to counterbalance trends towards European centralization in the face of grand societal challenges.*

Klüver, Lars, Rasmus Øjvind Nielsen, and Marie Louise Jørgensen, eds. *Policy-Oriented Technology Assessment Across Europe: Expanding Capacities*. Basingstoke: Palgrave Macmillan, 2016. DOI: 10.1057/9781137561725.0007.

European societies are pushed and pulled by tremendous forces in many directions at once. Science and technology has provided Europe and the world with incredible advances in production, health care, communication and almost any other aspect of human life. But our resource-hungry systems of production and consumption strain the supporting capacities of natural ecosystems, and we are moving from an era of abundance to one of scarcity. In a world of globally interconnected economies, systemic risks seem to increase exponentially and to far surpass the traditional managing capacities of nation states. Too often, however, the backbone reaction of decision makers is to invoke protocols of crisis management: gathering control in governmental centres and placing authority in the hands of narrow elites. In the case of the science-society relationship, large-scale research and innovation efforts accompanied by centralized social engineering regains prominence as decision makers attempt to take effective action. But while better knowledge and smarter solutions must undoubtedly be part of Europe's way forward, centralization in itself presents a danger to the social fabric of societies. Whenever societal decision making is disconnected from the perspectives of those that feel its consequences in their daily lives, alienation and dissatisfaction enters the relationship between governments and citizens. Attempts to address the grand societal challenges of our time must therefore first face the necessity of building capacities for effective democratic governance. Each step towards stronger centralized capacities for action must be accompanied by equal steps to build capacities for problematizing evidence, debating values and adapting solutions to fit local needs and cultural contexts.

The core message of this book is that technology assessment holds at least some of the needed answers for how we can build such decentralized capacities for knowledge-based democratic decision making. Technology assessment (TA) is a discipline of public administration that seeks to build bridges between research and innovation, society at large and political decision makers. To operationalize this institutional mission, a wide range of methods have been developed that enable TA organizations to dynamically address different gaps of knowledge and communication in different societal situations. As such, TA may be viewed as an institutional answer to the problem of governing research and innovation responsibly, where the problem of governance is seen first and foremost as a problem of decoupling between the different kinds of knowledge and different sets of values held by different societal actors.

DOI: 10.1057/9781137561725.0007

The emergence of a diverse policy support function

TA began as an interdisciplinary academic endeavour in the 1960s, at which time the long-term risks from indiscriminate use of modern chemical and nuclear technologies was becoming increasingly clear. Pitted against an establishment unwilling to admit to its own errors of judgement, TA first took the form of 'reactive' movement within academia, aiming to provide alternative evidence to support advocacy of mainly environmental protection and work-place conditions. This 'watchdog' role was expanded institutionally in the US at the national level when Congress established its Office of Technology Assessment (OTA) in 1972. Here TA – or parliamentary TA (PTA) as it came to be known – would act as analytical support to congressional oversight of the societal opportunities and consequences of the technological development.

TA was one of several strands of new, interdisciplinary forms of analysis seeking to provide guidance for decision makers in advanced industrial societies. Environmental impact assessment, risk assessment, foresight studies, technology ethics and the cross-disciplinary field of science-and-technology-studies (STS) all have their historical roots and institutional raison d'être in the apparent complexity of governing modern technology and the loss of popular trust suffered by experts and industrial stakeholders. There are many overlaps between these traditions in terms both of pragmatics of method and outlooks regarding the science-society relationship. The lines between TA and non-TA are thus not sharply drawn, and the different traditions mentioned continue to enrich each other.

In Europe, the first proposals for establishing capacities similar to that of the OTA were made immediately after the first round of European expansion in 1973. The idea of a common European Office of Technology Assessment, however, proved difficult for the member states to swallow, and a centralized unit dedicated to technology assessment and foresight would not see the light of day until the establishment of the Institute of Prospective Technology Studies (IPTS) as a subunit of the Joint Research Centre in 1992. Meanwhile, the TA concept had more immediate rapport with the individual national parliaments in Western Europe and the European Parliament itself. Beginning in the early 1980s and inspired by processes of knowledge sharing within the Commission-driven FAST program, TA institutions were established in connection with parliaments in Denmark, France,

DOI: 10.1057/9781137561725.0007

Germany, the Netherlands and the United Kingdom. An office for Science and Technology Options Assessment (STOA) was set up in connection with the European Parliament in 1987. At later stages, PTA organizations were also established in Belgium, Finland, Greece, Italy, Norway, Spain, Sweden and Switzerland while TA organizations, but without the 'P', established in Austria, the Czech Republic and within the Council of Europe have also been part of the landscape of TA in Europe (see also Chapter 1 and Chapter 2). Over the years, these TA institutions developed more 'proactive' roles for TA in supporting policy development. TA became closely linked with foresight studies and now shares the attempts to identify desirable pathways for development through forward-looking exercises. Some TA institutions took part in developing 'constructive' TA approaches to embed reflection on ethical, legal and social aspects (ELSA) in the development process itself. Other institutions developed methods for citizen participation and stakeholder inclusion in policy development for technological innovation and planning, precipitating the 'deliberative turn' in research and innovation policy. Today, TA thus walks on two legs: policy analysis and public engagement.

Since at least the turn of the millennium, the stakes of science and technology policy have been raised significantly. The perspectives of impending climate change and peak oil, which have been accompanied by increasing global competition in innovation, have driven science and technology policy towards more complex forms of reflexive governance. In this situation, the European TA field has increasingly sought to consolidate its methods and to provide 'strategic intelligence' for European policy makers acting at both national and European levels. The European Parliamentary Technology Assessment (EPTA) network was established in 1990 to enable cooperation among dedicated parliamentary TA units and units with similar goals. The IPTS has increasingly sought to orchestrate deliberation at European level between different TA and foresight organizations. And various parliamentary and non-parliamentary TA organizations have been increasingly involved in the European Commission's framework program for research, especially under those lines of research which are today known as Science with and for Society (SWAFS).

DOI: 10.1057/9781137561725.0007

Mobilizing TA for grand challenges – the PACITA Project

The PACITA (Parliaments and Civil Society in Technology Assessment) project was set up under the 7th European Framework Program for research and development. It ran from 2011 to 2015 and was coordinated by the Danish Board of Technology. Working under the assumption that TA will need to adapt to the change towards the internationalization of science, technology and policy, the project's overarching goal was to mobilize and expand the European TA community through processes of mutual experimentation and learning. Through such expansion, the working hypothesis of the project was that the TA field can grow into a Europe-wide support system for broadening the knowledge base of policy making in Europe. Helping to spread nationally based arrangements for providing TA services across Europe would serve the triple purpose of supporting national parliaments and governments, supporting and connecting national democracies across Europe in transnational dialogue and collaboration and helping to strengthen the bottom-up dimension of European democratic governance. We call this distributed support system 'cross-European TA'.

The PACITA strategy for an expanding TA field was bound up with a strengthening of national democratic institutions. In the four-year course of the project, it gathered a group of fifteen partner organizations from different European countries in collaborative processes, which were at once linked to European agendas and based on national debates. Among these partners, some are established TA organizations connected to parliaments or otherwise formally organized to support national policy (the partners from Austria, Belgium (Flanders), Denmark, Germany, the Netherlands, Norway, Spain and Switzerland), while others are organizations with closely related missions interested in developing locally appropriate institutional models for TA (the partners from Belgium (Wallonia), Bulgaria, the Czech Republic, Hungary, Ireland and Portugal). Among the members of this group, enough diversity with regard to national settings was represented that the outcome of the project would be applicable across EU28 and the group of associated or candidate countries.

DOI: 10.1057/9781137561725.0007

Main findings of the project

The project pursued four operational aims, the outcomes of which are documented in this book. The first aim was to map and conceptually categorize existing PTA institutions and practices. The second aim was to help guide countries, which as yet had no such dedicated TA functions, in establishing TA institutions appropriate for their specific culture and settings. The third goal was to showcase and give hands-on experiences with the praxis, methodologies, outcomes and social value of collaboration among TA institutions across Europe. Finally, the fourth goal was to begin the process of building up mutual capacities for training and communicating TA practice and results in order to build a cross-European TA capacity of infrastructures and human resources.

To what extend are the goals of such a project realistic? It is of key importance to assess the contributions as well as the limitations of what has been attempted. A key issue in this regard is the question of methods and how well they travel from their original national contexts to other cultures and to cross-national collaborations. For example, PACITA carried out a process of stakeholder deliberations on the future of ageing in which national responses were formulated to strategies developed at European level (see Chapter 7). Here, it was clear that the national processes in and of themselves were both politically useful and perceived as legitimate by the participants. And from a trans-European point of view, the simultaneous but nationally particular formation of ideas for policy presents a potentially highly valuable addition to the general European policy-formation process. But we must acknowledge at the same time that without a clear institutional mandate within the overall process of European policy formation, the recommendations produced by such nationally based bottom-up processes risk drowning in the whirlwind of European debates. Similarly, institutional issues produced profound challenges to an experiment in which the PACITA partners orchestrated a cross-national Future Panel. The Future Panel is a process in which parliamentarians from across the political spectrum take part in a common process of learning and forming opinions about complex issues that arise from science and technology. In this case, a cross-national panel would gather to learn about and debate the possible contributions of advanced genomics research to public health care in the future (see Chapter 6). Here again, while those parliamentarians

DOI: 10.1057/9781137561725.0007

who did take part were positively surprised and enthused by the spirit of deliberative inquiry that is embodied in the Future Panel, it was the lack of a common mandate from the involved parliaments which proved to be a stumbling block in the recruitment of parliamentarians for participation.

The benefits, however, should not be underplayed. A third PACITA experiment focused on public engagement and gathered citizens in different European member states in citizen summits to deliberate on the complex trade-offs involved in policy for sustainable consumption (see Chapter 8). This experiment provided strong indications that, when applied to the cross-European level, a deliberative take on public engagement seems to be a viable strategy for squaring the circle of democratic involvement in centralized European policy making. Simultaneous national processes in which citizens are briefed on the best available knowledge and afforded time to deliberate in socially diverse groups provides high-quality, nationally founded, but still 'European' inputs to the European policy process.

Looking at the method dimensions, the PACITA model of bottom-up development of cross-European TA that organized and operationalized by existing and emerging TA institutions and that was supported by the European Commission seems to be a viable pathway for sowing the seeds of cross-European TA. The outcomes of the project are surely tangible and promising. But at the same time, the PACITA project covered only fifteen countries and, as such, was an experiment, though a successful one. Ultimately, the idea of a Europe-wide implementation of TA must be taken up politically and given a mandate in order for cross-European applications of TA methods to really work.

The PACITA project may be said to have expanded European TA capacities in at least four different dimensions:

Geographically: We have aimed at expanding the capacity and formal institutionalization across Europe and have succeeded in doing that perhaps more importantly, we have also sown seeds for further expansion in the future.

Collaboratively: Developing cross-European TA for the benefit of Europe as well as for the member states has been a core aim of PACITA, and we have definitely proved that there is a large need for this and that there are big potentials in developing a truly European collaborative space for TA.

DOI: 10.1057/9781137561725.0007

Conceptually: The background, context and function of existing TA institutes have been scrutinized intensely with important new insights in the role and function of TA as a result.

Conceptually: The background, context and function of existing TA institutes have been scrutinized intensely with important new insights in the role and function of TA as a result.

Politically: At two parliamentary meetings with representatives from the EPTA and PACITA countries and beyond, it was clearly stated by the MPs that TA has a very important role to play for EU, Europe in a wider sense, and for the EU member states. A clear call has been made for a strong Commission engagement in widening the TA landscape in Europe and in providing options for new countries to take up TA.

Why cross-European TA?

What this book substantiates is the claim that going forward, the issue is not whether cross-European TA is possible. The book shows that the needed professional approaches exist, the national capacities can be built – often on the shoulders of existing ones – and collaboration between institutions distributed across Europe can be brought to work. Rather, the question is whether and why European policy makers and parliamentarians at national and transnational levels ought to support a vision of the development of cross-European TA capacities. The remainder of this introduction is dedicated to providing a frame in which to answer this question.

To begin with, we should try to get at the overall question whether there is in fact a need to strengthen national level capacities for policy analysis and public deliberation. The standard counterargument is that with global challenges we need global solutions and a strengthening of transnational decision-making capacities. For many, a 'return' to the nation state is unrealistic and represents in any case a step backwards. We are, however, not arguing for a 'return' to the nation state and a purely intergovernmental mode of European collaboration. On the contrary, our argument is that national democracies need strengthening in order to take their proper place in European – or global – multilevel governance.

The cornerstone of European collaboration remains the subsidiarity principle. And while the future will tell whether European collaboration will grow into federation, the sign of the times do not point in

DOI: 10.1057/9781137561725.0007

that direction. A realistic approach to drawing on Europe's collective strengths to efficiently address grand challenges must therefore take seriously the continuing role of national member states as a crucial level of policy adaption to local contexts. By the same token, with the process of European integration halted somewhere between inter-governmentalism and federated statehood, European institutions remain systematically under-democratized. Consequently, the national parliaments will remain privileged as fora for maintaining true European democracy.

What remains true logically, however, is challenged in real life. At the national level, the capacities of parliaments to act as counterweight to national executives have been systematically weakened by European integration. Parliaments have less formal access to providing input to common European policy processes than do governments. Parliaments therefore end up on the receiving end of the policy process, and the diversity of input that they represent is narrowed significantly. A similar effect of narrowing democratic diversity can be traced in the representative function of political parties. Here, the process of European integration has led the major centre parties in each member state to crowd around common middle positions compatible enough with the European mainstream to be strategically viable.

This process of consensus-building that centres on 'necessary' rather than 'wished' policies is amplified by the national economic strategic idea of the 'competition state' – the conception that international competition forces nations to act as if they were large companies. This conception has had highly detrimental effects on the range of futures and policies being imagined, and politically, it has inhibited the agility of centre parties, thus weakening parliamentary collaboration across parties.

The need for strengthening national parliaments is not about strengthening individual nations against the European community, nor is it a call for dis-integration. Rather, it is a call for strengthening precisely the part of the European system, which must be strong if Europe is to become legitimate

The role of cross-European TA in European governance

The inadequacies of the national and regional levels of governance are well understood and lie at the heart of the motivation for the development

DOI: 10.1057/9781137561725.0007

of the European Union in the first place. The capacities needed to continuously modernize society's infrastructure through research and development have long surpassed the size of the purses of individual nations. Consequently, cooperation on the advancement of research was one of the very first issue areas where the logic of cooperation became clear to European member states. Likewise, the scale of the mechanisms needed to render innovation economically viable has outgrown national markets. This is why the common European market has been a central guiding star for Europe for more than a generation and the European Research Area has such a prominent position.

With the pooling of innovative resources and merging of markets, much of the regulatory ability of member states has also shifted to the European level. One the one hand, this has allowed Europe to build a global region protected by the most progressive environmental and social protections in the world. But on the other hand, along with global deregulation to enhance trade flow, this shift has contributed to a lock-in situation for member states where increased cooperation is often not an option but the only viable path. Single member states are at a great disadvantage in relation to globalized industries and financial actors able to move production and capital from one country to the next. Countries wishing to move environmental and social policy forward – tools that will likely prove crucial in addressing grand challenges – are often bound to negotiate such changes within the traditional framework of European decision making known as the 'community method'. This is the framework in which national executives gathered as the Council of Ministers set out policy goals, which are then fleshed out in regulatory proposals by the European Commission to be approved by the European Parliament and ultimately the Council itself.

Often cited democratic dilemmas and deficiencies of the community method have led to the formulation of alternative governance strategies. The European Commission, for instance, has increasingly made use of soft governance approaches to coordinate societal actors around common goals. We see this in the response of the Commission to the Lund Declaration in the Europe 2020 strategy. Here public-private partnerships and networking initiatives meant to stimulate self-governance within industry are combined with a focus on societally strategic research and innovation. A cross-cutting framework to structure the self-governance of actors that participate in these strategic exercises is emerging under the title 'responsible research and innovation' (RRI). Within this framework,

DOI: 10.1057/9781137561725.0007

participation in research and innovation activities funded or otherwise stimulated by the Commission will be dependent on the willingness to undertake self-governance measures to align R&I output with the needs of society. Such measures, whichever practical form they may take, shall enact the principles of inclusiveness, anticipation, reflexivity and responsiveness. The ideal embedded in these principles is those of a self-regulatory system of multiple societal actors able to converge on common goals through ongoing dialogue and mutual learning. To a very large extent, this ideal has always been shared by the technology assessment community. Whether TA has been reactive, proactive, 'constructive' or 'participatory', TA has always sought to embed upstream societal reflections in the real-world processes of science, technology and innovation policy, precisely to achieve outcomes that would be already well rounded and aligned with the needs and values of multiple societal actors. The only major point at which the TA project still stands out from the RRI framework – and the point around which the unique value TA may add to RRI crystalizes – is the practical and institutional commitment to retain and strengthen the embedding of such soft governance approaches in the institutions of representative democracy.

Cross-European TA – still in the sense of national policy-oriented TA bodies in all states collaborating at European level – may thus play a number of important roles in consolidating the ideas in modern European governance:

▸ The need to strengthen national parliaments in the EU is broadly acknowledged, but the structures to facilitate that change are lacking. Here, TA can play an important role by serving parliaments with knowledge, analysis and debate on EU developments in science, technology and innovation.
▸ The importance of the subsidiarity principle is greater than ever, but adhering to it may produce locked decision-making situations under the community method. Circumventing such dead ends demands the creation of spaces for open explorative dialogue across the EU, involving citizens, stakeholders and parliaments. TA has longstanding traditions which make it an obvious player for creating such cross-European analytical dialogue.
▸ Governments are forced to become more and more European, while parliaments become increasingly national – some may even say provincial. TA can build bridges for parliaments across Europe,

DOI: 10.1057/9781137561725.0007

thereby enhancing the connection between parliamentary debates and European developments.

▸ The EU needs to get in contact with citizens and to support the emergence of a true 'European public', but it faces a lack of European identity. With national TA institutions in place, a platform emerges with the legitimacy to engage and consult citizens on the national level and connect the outcomes at the EU level – which makes TA a potentially perfect partner for both the national and the EU level governance.

▸ Cross-European TA collaboration can add to the smart specialization aims by, on the one hand, facilitating the needed discourse at the national level and, on the other hand, ensuring that it is connected across Europe – allowing for a certain level of coordination of the specialization.

▸ TA at the national level is an important factor for having a rich analysis and conversation about the societal opportunities and challenges stemming from science, technology and innovation. Having TA institutionalized in all European states will provide an opportunity for expanding that analysis and conversation to the European level and creating much needed links between the multiple levels of the European governance system.

The PACITA consortium has on the basis of these thoughts and the lessons of the PACITA project provided the TA Manifesto, which has gained support from more than 300 signatories.[1]

Note

1   See http://www.pacitaproject.eu/ta-manifesto/.

OPEN

The TA Manifesto

Abstract: *A common statement from the organizations involved in the PACITA project, the PACITA manifesto argues for the necessity of European political support of future efforts to expand technology assessment (TA) capacities in the European member states. The authors posit the tradition of technology assessment in European as a democratic project to inform policy makers on societal and environmental topics related to science, technology and innovation. And they call attention to the necessity of countering the increasing influence of science and technology on societal development and policy making with increasing capacities for technology assessment. Developing a more comprehensive 'policy-oriented' approach to TA is called for by the authors along with an increase in cross-European collaboration in TA.*

Klüver, Lars, Rasmus Øjvind Nielsen, and Marie Louise Jørgensen, eds. *Policy-Oriented Technology Assessment Across Europe: Expanding Capacities.* Basingstoke: Palgrave Macmillan, 2016. DOI: 10.1057/9781137561725.0008.

Expanding knowledge-based policy making on science, technology and innovation

Technology is a central element in the policy response to the great challenges of our time, such as ageing societies, climate change and public health. In addition, emerging technologies such as synthetic biology, nanotechnology and the ever-changing Internet all challenge established policies. The encompassing quality of technology today is influencing the lives of citizens all over the world. The global transforming power of technology, thus, has to be aligned with policy making and democracy.

Technology assessment (TA) can be seen as a democratic project in Europe, providing and supporting robust and knowledge-based policy making on societal topics related to science, technology and innovation. It has mostly been established in the western parts of Europe and in connection to national parliaments.

Technology development and policies are becoming transnational. At the same time, the need for multilevel action on the grand challenges of our societies is obvious. Modern policy making needs to bridge these transnational and multilevel dimensions of the development, regulation, implementation and management of technology. The rapid technological development, in combination with science and technology's profound influence on societal developments and policy making, call for an important and increasing role for European TA in the future.

The PACITA project has during 2011–15 enhanced European TA by:

▸ enhancing the capacity for doing TA in and across European nations;
▸ increasing cross-European collaboration in TA;
▸ expanding the institutionalization of TA across Europe;
▸ developing the conceptual framework of TA into a more comprehensive 'policy-oriented approach', adding to the traditional parliamentary-oriented TA in Europe;
▸ raising awareness about the possibilities for modern policymaking that lies in TA.

TA – a multi-level and cross-border European capacity for the future

The PACITA project should be seen as a new setoff for a necessary expansion of the European TA landscape:

DOI: 10.1057/9781137561725.0008

TA should collaborate to increase the capacity of providing robust and independent policy advice for policy makers in all of Europe. As the EU grows and Europe becomes more connected, TA can through strong knowledge sharing and collaboration contribute to knowledge exchange and synergies, which provide for widespread use of the independent and knowledge-based advice from TA. Countries should help each other by sharing TA knowledge and outcomes.

TA should be institutionalized in all European countries in order to provide for independent knowledge-based advice and to promote the engagement of stakeholders, experts, citizens and policy makers in a collaborative, democratic provision of policy options. The diversity in cultures and political contexts in Europe call for national implementation of TA in ways which are optimal for the single nation. For Europe to develop strong knowledge-based and democratic decision making, TA needs to be implemented in all European states.

There is a clear political call for increased parliamentary dialogue across Europe on the technological development and its meaning for our societies. TA should play an active role in setting up that dialogue. In a context of globalization and European construction, policy making on many science- and technology-related issues needs a cross-border approach. As stated by two parliamentary meetings in PACITA, TA has an important role to play in setting up parliamentary dialogue across Europe.

Citizens in Europe have a democratic right to be heard about the technological development since technology is strongly influencing their lives. PACITA has proven that TA has the methodology to make that right happen on the European level. Over the years, TA has developed a toolbox of methods and approaches for engaging different groups of actors, and especially the involvement of citizens in policy debates. Seeing that the 'grand challenges' will demand an understanding of scientific and technological analysis as well as of societal values, TA is well suited to giving advice on these topics, also based upon citizen engagement.

Strong TA collaboration on the project level across Europe should be encouraged and supported. The development of technology moves forward with increasing pace. Because these developments happen on a European and international level, the need for cross-European TA is evident. Collaboration between countries and institutions will ensure that knowledge from experienced TA units is combined with new thoughts and ideas from emerging TA actors.

DOI: 10.1057/9781137561725.0008

TA has a crucial role to play in the European strive for ensuring societally responsible research and innovation. Responsible Research and Innovation (RRI) has shaped the recent years' policy discourse in Europe related to the societal role of research and innovation. It has given greater focus to key concepts in TA, such as participation, forward-thinking, reflexivity and policy action. TA can and should be a key carrier of the concept and should play a light-house role in RRI.

DOI: 10.1057/9781137561725.0008

Part I
Expanding Technology Assessment

▶

DOI: 10.1057/9781137561725.0009

OPEN

1

Seeing Technology Assessment with New Eyes

Rinie van Est, Michael Nentwich, Jurgen Ganzevles and André Krom

Abstract: *Van Est et al. present a 'relational' model for analysing technology assessment (TA) institutions. Expanding on metaphor of TA as a bridge between science, society and policy, the authors describe how such bridges are established in terms of network relations. European TA institutions in various ways link parliaments and governments with civil society and science. In part, TA projects provide such linkages, but importantly, TA institutions in themselves also provide informal personal links between societal spheres. With in-depth examples from different European member states, Van Est et al. provide institutional entrepreneurs with rich material for imagining institutional TA arrangements that might fit within their own national arenas.*

Klüver, Lars, Rasmus Øjvind Nielsen, and Marie Louise Jørgensen, eds. *Policy-Oriented Technology Assessment Across Europe: Expanding Capacities*. Basingstoke: Palgrave Macmillan, 2016. DOI: 10.1057/9781137561725.0010.

DOI: 10.1057/9781137561725.0010

Creating institutional platforms for technology assessment (TA) has proved possible via different nationally specific pathways. In examining these pathways, previous reflections on the institutionalization of TA have focused on the relationships between TA institutions and national parliaments. However, movements both internal and external to TA mean that relations to other societal spheres have gained increasing importance for many TA institutions. In order to provide insight into the full range of possible institutional arrangements for delivering policy-oriented TA services, we provide a model for the network relations that help to create and sustain TA institutions. We then draw out implications for the design of S&T governance.

A relational framework allows for a better understanding of technology assessment and its role within the complex of institutional relations underpinning the governance of science and technology (S&T) in society. Understanding TA in relational terms implies taking full account of the position that TA occupies in a social network (e.g. a governance network) and acknowledging that various bonds enable and constrain the activities of organizations performing 'TA-like' functions. We apply this model to existing TA institutions and develop a typology of ways that TA may evidently fit within national institutional contexts. Our motivation is to help institutional entrepreneurs and political supporters of emerging TA platforms to imagine arrangements that will fit their specific national arenas. We seek to provide evidence of the relations between TA, other public institutions, and other societal sectors in order to guide strategic processes of network-building around the promotion of national TA capacities. Moreover, we argue that TA can and should be seen as a necessary part of democratic S&T governance.

The model expands upon a long-standing metaphor for TA as a provider of 'bridges between science, society and policy' (Decker and Ladikas, 2004). The model concretely maps the relationships between existing parliamentary technology assessment (PTA) institutions and four societal 'spheres' involved in S&T governance, namely parliaments, governments, S&T, and (civil) society. The mapping takes into account a range of mechanisms of interaction between these spheres, distributed on a macro (institutional), meso (organizational) and micro (project) levels. The model thereby illustrates how (P)TA functions in terms of information exchange and relational trust-building between different societal actors.

Comparing the results of our case studies, it is clear that 'parliamentary TA' is much broader than the label suggests. While parliament

DOI: 10.1057/9781137561725.0010

remains an essential base for most existing policy-oriented TA organizations, building and maintaining credibility towards actors within government, S&T, and society in the broad sense is important for operating effectively and with legitimacy – even for TA offices nested inside parliaments. Five different organizational variants of TA are currently operational where different weight is given to each of these societal spheres. There are thus many strategies to pursue in countries that want to establish TA-like support functions, and the material provided here will help to make the best of the opportunity structures that exist in each individual country.

Lessons learned, relevant to promoters of TA-like arrangements, include:

▶ Acknowledge the dependence of TA in order to achieve independent advice with an impact
▶ Consider the whole institutional possibility space when setting up new TA organizations
▶ Foster relationships on the institutional, organizational, and project levels

Background

Throughout its history, three concerns have been of fundamental importance to the practice of PTA, namely:

▶ how to institutionalize PTA
▶ how to structure PTA organizations
▶ how to design and perform PTA projects

For example, the establishment of the Office of Technology Assessment (OTA) in 1972 in the United States presented a real institutional innovation. OTA was meant to provide Congress with 'unbiased' information concerning, for example, the social and political effects of technologies. The establishment of a congressional TA bureau was a way to redress the imbalance between legislature and executive with regard to technological change, and thus it was an attempt to strengthen the representative model of democracy (Van Est and Brom, 2012). When during the 1980s several European countries created PTA institutions, the focus was also quite naturally on institutionalizing and organizing PTA. A key issue in

DOI: 10.1057/9781137561725.0010

this debate was how the relationship between the Parliament and the TA organization should be shaped to make it fit comfortably in the specific political cultures of each country.

In some countries, such as Denmark and the Netherlands, controversies over technologies were seen not only as a matter of power balance between the government and each parliament but also as a problem between the government, the parliament, and the wider public (Van Eijndhoven, 1997). As a result, in these countries public education and debate were seen as central to the mission of PTA, which led to early experiments in 'participatory' TA. In the 1990s, growing uncertainty and societal disagreements concerning pathways for technological innovation and economic development led to increased political interest in the use of participatory methods to achieve legitimacy of hard political choices that were made in situations where science could provide only soft evidence, and these choices would need legitimacy through public deliberation and consent (Funtowicz and Ravetz, 1992). During this period, debates in the PTA community (facilitated for instance by the EUROPTA project) sought to consolidate practical experiences with public engagement and to arrive at mutual understandings of how to design and perform participatory TA projects (Joss and Belluci, 2002) – for instance, the role of project management, the choice of methods (Van Eijndhoven and Van Est, 2002), and the impact of participatory TA (Hennen, 2002).

At the turn of the millennium, however, the initial wave of 'participation optimism' at the political level was countered by demands for evaluative evidence of the positive effects of linking citizens' participation and stakeholder dialogues to processes of policy formation based on expert input. To maintain its political legitimacy and mandate, the PTA community thus became concerned with the visibility and impact of its own activities. In the TAMI project (Decker and Ladikas, 2004), this led to a wider reflection on the types of impacts that TA processes could have on different clients in different situations and how the institutional context of a PTA organization served to both enable and constrain the impact that TA could have on various publics (Cruz Castro and Sanz-Menéndez, 2005). Reflections on the practicalities of achieving impact in a world of distributed network communication led the TA community to focus on multiplatform communication (policy briefs, personal networking, websites, blogs, and media appearances).

DOI: 10.1057/9781137561725.0010

The compounded output of these debates can all be traced in the so-called process definition of TA, which became standard after the TAMI project:

> Technology assessment is a scientific, interactive and communicative process which aims to contribute to the formation of public and political opinion on societal aspects of science and technology. (Bütschi et al., 2004: 14)

Today, we see a need to articulate the relevance of approaches to policy support developed within TA in a new and broader context of grand societal challenges. Here there may be a need for 'non-PTA' actors to take up and carry on the same practices. To this end, the openness of the definition of TA inherited from the TAMI project allows us today to apply the definition to a much broader field of organizations that work to provide similar forms of support to decision makers involved in S&T governance. The framework presented here can be used to clarify the institutional roles that various forms of TA or TA-like organizations can play within the governance of S&T.

The framework: TA understood in informational and relational terms

TA can be described in both informational and relational terms. On the one hand, the informational view characterizes TA practices based on the particular knowledge that they generate, namely knowledge about the societal aspects of S&T. The relational approach, on the other hand, starts with the insight that the TA field owes its continuing existence and position to support from its clientele. Our framework combines the two approaches based on the understanding that the informational and relational aspects go hand in hand. In support of this framework and adding to existing knowledge on TA, we try in the following first of all to come to grips with the relational aspects of TA.

Modelling TA in relational terms

Understanding TA in relational terms implies taking full account of the position that TA occupies in a social network (e.g. a governance network at regional, national, or European level) and acknowledging that various bonds enable and constrain the activities of organizations performing 'TA-like' functions. To create an evidence base for analysing

DOI: 10.1057/9781137561725.0010

these relational factors, we scrutinized the interaction between existing PTA organizations and various social actors (Van Est and Ganzevles, 2012, Ganzevles et al., 2014, PACITA, 2014). The following four societal 'spheres' were defined to group actors in the institutional landscape around PTA organizations: parliament, government, civil society, and S&T. The choice of these four spheres was dictated by the most common characteristics of European PTA. For PTA organizations, their institutional linkage with parliament is of primary importance. Government, however, may also play a crucial role – for example, as a sponsor but also a recipient of advice. In addition, relationships with civil society (in the case of public participatory TA) may play an important role in the practice of PTA. And since PTA is ultimately about governing S&T, the model could not have done without the inclusion of S&T as a societal sphere. Of course, these choices do not imply in any way that other spheres such as media, industry and business are not relevant in many ways to TA in general.

To map existing models in terms of their relations with the four selected societal spheres, PTA organizations were asked to express the involvement of each of the four social in percentages. The results show that PTA organizations indeed establish and maintain multiple relationships with the four discerned social spheres. PTA organizations differ from each other to the extent that they interact (on the institutional, organizational, and project levels) with the four distinct social spheres. Out of the fifteen theoretically conceivable interaction models, the mapping process in the PACITA project identified five distinct PTA models that are currently operational in Europe.

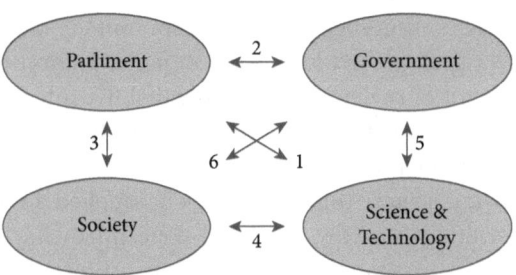

FIGURE 1.1   *Four spheres involved in the relation model of PTA*

DOI: 10.1057/9781137561725.0010

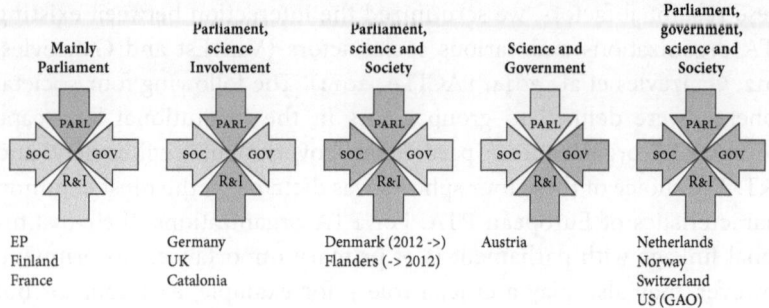

FIGURE 1.2   *Currently operational models of (P)TA*

We studied the linkages between TA and the four distinguished social spheres on three (interconnected) levels: institutional, organizational, and project. The macro-level or institutional-level concerns the political support for a TA organization that has the parliament as one of its main (formal) clients. It also concerns the way in which TA is legitimized and framed as an institutional solution for the governance of, often societally controversial, developments in research and innovation. The meso-level, or organizational level, concerns the politics of shaping and controlling the TA organization that has the task to perform PTA. Finally, the micro-level, or project level, relates to doing TA. Issues at this level are as follows: choices about the framing of the topic, choices between kinds of method, and strategies for establishing communication between the project and parliament or other recipients.. Our modelling of TA in relational terms is founded on the notion of informational interaction mechanisms, loosely defined as communicative procedures or routines on the institutional, organizational, and project levels for enabling and constraining the involvement of actors from the above-mentioned four social spheres in shaping the practice of TA. We discern nine interaction mechanisms: client, funding, evaluation committee, board, working program, project staff, project team, project participatory methods, and project revising and/or reviewing. While the first five interaction mechanisms play out on the institutional and/or organizational levels, the latter four all play out on the project level.

In the following pages, this framework is applied to three different cases, illustrating how the relational conceptualizing of TA(-like) activities may help to analyse the process of institutional pathfinding and adjustment, as well as institutional issues that underlie concrete TA projects.

DOI: 10.1057/9781137561725.0010

Case 1: pathfinding in Bulgaria

The relational model of TA can also be used to make emerging developments explicit, pointing to still-fragile structures, providing a snapshot
of where a country is on a potential evolutionary pathway for TA. We
use the case of Bulgaria to illustrate this.[1]

The TA-landscape

Bulgaria is in a highly explorative phase when it comes to dealing with the
societal issues of S&T. The PACITA-partner, ARC-FUND, is a central player
in this field. Its task is to 'shape policies and developments towards information society and knowledge economy in a national, regional, European
and global context'. The national Academy of Sciences is another important
actor. In Bulgaria, expert advice (like TA) to policy makers is a delicate
matter. Besides a high level of public distrust in the political system, recent
years also show an erosion of trust in scientific institutions. This creates a
vicious circle in which policymakers rarely ask for expert advice and policy
making is perceived as lacking a sufficient knowledge base.

In 2012 a temporary parliamentary committee on shale gas was set up
to carry out activities, which – from a TA point of view – resemble a
PTA project. The committee had some months to study and discuss good
practices and legislative options for the environmentally safe exploration
and mining of shale gas. Three hearings with external experts were held.
MPs in the committee mainly listened; some complained; and others
seemed to feel offended by the views of the experts. Both actors from
the realm of S&T and representatives of NGOs were invited. These
activities could have been a good starting point for setting up more of
these PTA-like activities since a good example tends to be followed. The
committee, however, has been subject to strong criticism: its objectivity
and impartiality were doubted. It seems that objective, multidisciplinary
analysis, interpretation, integration, and review of the knowledge gathered in the hearings were lacking. Developing TA-like skills and capacity
might help make such TA-like activities trustworthy from both a political and a societal point of view.

A government – society – S&T network forum

The PACITA project enabled ARC-FUND to search for organizational
and institutional TA-capacity. For several reasons, ARC-FUND considers

DOI: 10.1057/9781137561725.0010

the governmental branch a more favourable client and sponsor of TA than it considers parliament: to a large extent, the government branch governs the political decision-making process; preoccupied by the next election, politicians have little interest in 'long-term', complex S&T issues; the government has adopted a new national innovation strategy, to which the early 'horizon scanning' of societal issues, related to S&T developments, can contribute.

ARC-FUND's institutional strategy is to act as a network secretariat ('staff' in our modelling) for TA-like activities in Bulgaria. The formation of a cross-disciplinary TA network is aimed for, in which representatives of expert-based organizations, think tanks, and policy institutions are represented (board, committee, panel, or platform in our TA model). ARC-FUND aims to increase both awareness about TA as well as the level of societal debate (relevant for the 'client' category in our modelling). A TA network forum is foreseen, gathering annually for a public debate on the most pressing S&T related issues (cf. 'working program' in our model). There is no guarantee that this will lead to a formal institutionalization of TA. But various actors have addressed the need for a pilot project in order to 'prove' the relevance of TA for Bulgaria – preferably within the relevant organizational and institutional structures.

Case 2: Institutional re-adjustment in Austria

The relational modelling of PTA institutions enables us to map dynamic developments of existing organizations as relations change over time. Political dynamics may result in the shifting importance of the four societal spheres, to which the organization relates itself. One current case of such 'drifts in the possibility space' is Austria. Since the ITA is deeply rooted in the academic world and has a high proportion of studies carried out for government, the Austrian situation can be described as 'shared science-government involvement in TA'. Lately, however, we observe a slow move towards 'shared parliament-government-science-society involvement in TA' in that both the national and European parliaments are becoming more important as clients for the ITA just as the citizens become active participants in projects and target groups for increasing public-relations activities.

DOI: 10.1057/9781137561725.0010

Strengthening connections with society and parliament

First, Austria's core TA organization, the Institute of Technology Assessment (ITA), has expanded its portfolio considerably towards greater involvement of society. One the one hand, participatory procedures are gaining importance in the ITA's work programme and are at the centre of many ITA projects. While a few years ago the ITA mainly observed the developing participatory TA approaches, contributed to theoretical projects such as EUROpTA, or assessed participatory events carried out by others, the ITA is now involving citizens and stakeholders on a regular basis. On the other hand, its mother institution, the Austrian Academy of Sciences – as well as the Federal Ministry of Science, Research and Economy – push the ITA towards an intensified relationship with society. As a consequence, a professional public-relations unit has been set up inside the institute, not only feeding the new Internet-based social media but also playing an growing role in the ITA's public events and project dissemination activities.

Second, while there has been only limited contact between the ITA and the Austrian Parliament ('Nationalrat') for almost two decades, the situation has been changing since 2012. The Nationalrat has shown increased interest in TA. In particular, its Research, Technology and Innovation (RTI) Committee has invited the ITA on several occasions to present TA work and to explain what it could contribute to parliamentary work. The acknowledgement of technology assessment as a potentially valuable contribution culminated in 2013 with a full membership of EPTA. Since then, the ITA is in regular exchange with parliamentarians, offering amongst other things a newly devised policy-briefs series explicitly targeted towards MPs. These so-called ITA-Dossiers are two-pagers that present TA topics in plain language and with a focus on possible political action. Most recently, in mid 2014, the Nationalrat decided to solicit a study on how to best implement advice and input with regard to TA and foresight for the Austrian Parliament. This one-year study will produce concrete proposals for the future relationship between the Nationalrat and, in particular, the ITA. A pilot project on 'Industry 4.0' is also under way in 2015, with a view to include these experiences in the recommendations. For these projects, the ITA is partnering with an institute that specializes in foresight and technology policy, so the Austrian Parliament can be said to be knitting closer ties with the TA and foresight communities. Two further developments support this growing

DOI: 10.1057/9781137561725.0010

importance of the parliamentary level: first, the mother institution of the ITA, namely the Austrian Academy of Sciences, has started offering its competencies to the Nationalrat; presentations and debates of recent societally relevant research done in the Academy are planned as regular events in the premises of the Parliament. Second, the ITA became a member of the European TA Group (ETAG), carrying out projects for the Science and Technology Options Assessment (STOA) panel of the European Parliament. So far, four such projects were concluded.

Case 3: Placing a TA project in a cross-national context

The relational model can usefully be applied to concrete TA projects. The PACITA sub-project 'Future Panel on Public Health Genomics' had a transnational approach and involved a consortium of organizations from both PTA and non-PTA countries. It made use of the Future Panel method, in which, from the very start, a panel of MPs (the Future Panel) co-determines the research agenda, together with a broad range of experts and guided by TA specialists. In the PACITA experiment, the Future Panel method was used in a cross-European context. In this sense, the project was truly a methodological experiment (see Chapter 6).

Analysing this project at the micro (project) level, the meso (organizational) level, and the macro (institutional) level enables us to highlight some essential connections between these three levels and formulate some lessons for the future use of TA methods in a cross-national context. We learn that there is *therefore* a need for more knowledge about how the relational basis is established for TA in networks of organizations and on the transnational level.

At the project level, an important aim of the sub-project was to support evidence-based policy making on Public Health Genomics (PHG). However, it turned out to be difficult to connect the evidence base provided on a range of issues related to PHG to the European political and policy debate in a constructive way. The Future Panel consisted of MPs from different national parliaments, who had to discuss policy issues and options concerning PHG on a *European* level. Accordingly, the research and policy agenda that evolved in the PACITA project did not always match the political issues and the context, which members of the Future Panel, and members of the task team had to face on the national level. This gap between the national and European political agenda also

DOI: 10.1057/9781137561725.0010

limited the opportunities for dissemination of the project results, at both the European level and the national level.

At the organizational level, the close cooperation between established (P)TA institutions and organizations in countries without such institutions presented some practical challenges. These challenges, however, were taken into account to stimulate mutual learning and are discussed in Chapter 6. The cross-national dimension of these challenges, however, needs special attention. Within the PACITA project, the relational TA perspective was applied to clarify the interactions between *one* particular organization and the various identified social spheres: parliament, government, society, and S&T. But the team responsible for the Future Panel on PHG was not drawn from one organization with a clear position in the 'possibility space' of TA at the European level. In fact, the team was deliberately composed of members who represent organizations with *different* positions in that possibility space. There is a clear lack of knowledge about how TA projects are set up in cross-national networks of organizations.

At the institutional level, the institutional conditions for effectively connecting the project results to policy making were not in place. Future Panel members were invited as individual MPs, with no formal appointment by their respective parliaments. As a result, the connection between the project results and the respective parliaments was not very robust. And although funding was in place, it was not clear who the client actually was. We think that this is also true for many other FP7-funded projects. Many European Commission–instigated experiments revolve around the possibility of cross-European TA-like activities (Barland et al., 2012). One might argue that the EC is the client since it funds the projects and since EC-funded projects typically involve reporting in the form of sending deliverables with the project results to the EC. Our way of looking at TA presents a more involved type of client, either on the project, organizational, or institutional level. This raises the question of whether the proper institutional conditions are in place to truly connect the outcomes of EC-funded cross-European TA-like activities to policy making.

Lessons learned: Implications for the democratic governance of S&T

Defining TA in relational terms opens up a new way of understanding TA and leads to a new way of questioning TA and both its role and impact

DOI: 10.1057/9781137561725.0010

in the way that modern society deals with S&T. This section explores what implications our new approach has for the future of TA and, more generally, for the democratic governance of S&T. We believe that this set of lessons is relevant not only to the TA community but also to all kinds of TA-like activities, one important instance being the responsible research and innovation (RRI) activities that will be developed in the context of Horizon 2020.

The lessons learned are structured by the three key elements of our model: (1) connecting to four societal spheres; (2) making connections on the micro-, meso-, and macro-levels; and (3) making connection by means of interaction mechanisms. Our reflections have led to nine lessons.

TABLE 1.1   *Key elements of the relational model of TA and related research issues and lessons learned*

Key elements of the relational model of TA and related research issues	Lessons learned
Connecting to four social spheres	
• **Characterizing TA**	Lesson 1: Understanding TA in informational and relational terms is useful
	Lesson 2: TA can effectively play out in many institutional and organizational forms
• **Bridging PTA- and non-PTA-countries, and PTA and TA countries**	Lesson 3: Intellectual playing field needed between PTA, non-PTA and TA
	Lesson 4: When setting up new TA organizations consider the whole institutional possibility space
• **TA and the governance of S&T**	Lesson 5: Acknowledge the institutional and organizational constraints that the governance of S&T may face
• **Long-term institutional dynamics and adaptability**	Lesson 6: Existing TA organizations need to adapt to changing demands
Making connections on the micro-, meso- and macro-levels	
• **Making connections on three levels**	Lesson 7: Foster relationships on the institutional, organizational, and project levels
• **Organizational and institutional conditions for successful TA project**	Lesson 8: Improve organizational and institutional conditions for the success of TA-like activities
Understanding interaction mechanisms	Lesson 9: Acknowledge the dependence of TA organizations, in order to achieve independent advice with an impact

DOI: 10.1057/9781137561725.0010

Connecting to four spheres

Characterizing PTA

Research within the PACITA project shows that PTA organizations indeed establish and maintain multiple relationships with the four discerned social spheres. PTA organizations differ from each other to the extent that they interact (on the institutional, organizational, and project levels) with the four distinct social spheres. As we saw earlier, the mapping process in the PACITA project identified five distinct TA models that are currently operational in practice in the field of PTA. The PACITA research thus confirms that it makes sense – both conceptually as well as practically – to talk about PTA in terms of its relationship to four spheres – parliament, government, society, and S&T. Moreover, PTA can and does play out in many different forms, and these forms can all be effective in their own manner. Consequently, the following two lessons can be drawn:

Lesson 1: Understanding TA in informational and relational terms

From both a conceptual and a practical point of view, it is important to understand TA both in informational terms (as a form of science-based policy advice) and in relational terms. According to the relational view, it is essential to consider the relationships of knowledge sharing and trust that TA organizations build up and maintain with different societal spheres, such as parliament, government, society, and S&T.

Lesson 2: TA can effectively play out in many institutional and organizational forms

Each of the models identified in the study can be effective in a specific context.

Bridging PTA and non-PTA countries, and PTA and TA countries

Our model has been developed to characterize TA institutes. As a result, the model can be used to typify TA organizations that either do or do not have a parliament as one of their clients. This is illustrated by the Austrian TA organization ITA, which was characterized as 'shared government-science involvement in TA'. Our model thus creates an intellectual level playing field between PTA and TA organizations, and also between PTA

DOI: 10.1057/9781137561725.0010

and non-PTA countries, and even TA and non-TA countries. Creating such an intellectual level playing field has been a major drive behind the PACITA project because it is a necessary condition for mutual learning between PTA and non-PTA countries, which was the key objective of PACITA. Our inclusive model acknowledges the similarities between the various types of TA – ranging from parliamentary towards constructive TA and even non-institutionalized forms of TA – and enables us to study the similarities and differences between the various TA organizations and their activities. Based on this argument, we draw two further lessons:

Lesson 3: Intellectual level playing field is needed between PTA, non-PTA, and TA

The relational conception of TA creates an intellectual level playing field between PTA and non-PTA countries, between PTA and TA organizations, and treats various types of TA-like activities on an equal footing. This is a necessary condition for stimulating a mutual learning process between different countries, organizations, and TA-like activities. This perspective is also needed to show the added value of TA within the broader network of S&T governance activities.

Lesson 4: When setting up new TA organizations, consider the whole institutional possibility space

Since TA can play out in many different forms and since each can be effective in a specific context (see lesson 2), countries with an interest in setting up TA are encouraged to consider the whole 'possibility space' in order to select the model that is particularly suited to their political and societal demands and their institutional contexts.

TA and the governance of S&T

TA plays a role in the broader challenge of the democratic governance of S&T. Since our model treats various types of TA institutes and various types of TA-like activities on an equal footing, it opens up possibilities to study to what extent various TA institutes within a national or international setting can complement each other. In order to understand the complexities of the governance of S&T, there is a strong need to reflect on the interaction between the various research and engagement processes in the various social spheres and to reflect on the organizational and institutional constraints that these processes encounter. Such a

DOI: 10.1057/9781137561725.0010

comprehensive approach is especially needed to get to grips with the particular added value of TA within the broader national network of S&T governance activities.

Lesson 5: Acknowledge the organizational and institutional constraints that the governance of S&T may face

In order to understand the complexities of the governance of S&T, we need to reflect on the interaction between the various research and engagement processes in the various social spheres and to reflect on the organizational and institutional constraints that these processes encounter.

Long-term institutional dynamics and adaptability

Appreciating the dynamics of TA on the institutional level is crucial for the future of TA, with regard to creating new institutions and maintaining existing institutions or adapting them to new political demands. Our model makes it possible to study the institutional development of a TA organization over a long period of time. The PACITA project shows that we need to take into account a long-term perspective to get to grips with that process. For example, it was found that in many countries the political debate about setting up PTA took a long time, often more than a decade. Moreover, existing institutes may radically or gradually change their institutional position.

Lesson 6: TA institutes need to adapt to changing demands

Over a longer period of time, the political and societal demands for TA change. In order to survive, existing TA organizations have to adapt to these changing circumstances. The 'space of possibility' offers ample opportunities for such adaptation. For example, a country may first set up a TA organization and later on gradually develop its PTA capacity, by building up stronger relationships with parliament and include parliamentary TA types of activities.

Making connections on the micro-, meso-, and macro-levels

Our model stresses that the relationships between the TA organization and the various social spheres are developed and maintained on three levels, each of which has its specific features and dynamics. Up till now, most research efforts have been put towards understanding and mapping

DOI: 10.1057/9781137561725.0010

the relationship between PTA and parliament on the institutional level. The country reports of the PACITA project (PACITA 2012) is one of the first attempts to get to grips with how the relationship between PTA and the parliamentary process is shaped on the project level. Although these, often personal, contacts on the practical level often have a major effect on the impact of PTA, these types of activities of a PTA institution are rarely mapped or reflected upon. And how contacts between PTA and parliament are shaped on the organizational level is well known for PTA organizations that work very close with parliament, but they are far less known for the PTA organizations that operate at a distance to parliament. In addition, even less is known about the way in which PTA organizations set up and maintain relationships with the other three social spheres: government, S&T, and society. Here another complexity pops up in that these spheres consist of networks of organizations. It would be valuable to have more knowledge about to what extent and in what way a TA organization organizes and maintains its connections with various clusters of organizations (e.g. different governmental institutions.

Lesson 7: Foster relationships on the institutional, organizational, and project levels

Relationships between TA organizations and the various social spheres are developed and maintained on the institutional, organizational, and project levels. So far, literature on PTA institutions has focused on the institutional relationship between PTA organizations and parliaments, and too little attention has been given to the relationships of such organizations with the other social spheres and how contacts are shaped on the organizational and project levels.

Organizational and institutional conditions for successful TA projects

The description of TA methods often focuses on the project level. Our model implies that the impact of a certain method will also depend on institutional and organizational conditions. This dependency has received little attention from both scholars and policy makers. Most methodological descriptions take for granted that a TA organization with the proper human capacity and skills exists to perform the method and that such an organization has the proper institutional mandate

DOI: 10.1057/9781137561725.0010

to perform the method. This, however, is not the case, neither on the national nor on the international level.

An important question that will be addressed is: if a particular TA method developed at the national level is used on the European political level, then to what extent does the impact of that method depend on well-developed relationships between TA and the political system on an institutional and organizational level?

At the moment, the notion of responsible research and innovation (RRI) politically frames, enables, and constrains contemporary discourse on how to properly enact the democratic governance of innovation. In the context of Horizon 2020, many TA-like RRI activities will be sponsored and set up. Also, in this context, it is important to address not only methodological questions, but also questions about the organizational and institutional conditions needed to guarantee a proper impact of those activities.

Lesson 8: Improve the institutional and organizational conditions for success of TA-like activities

The policy impact of a certain TA method will depend not only on the quality of the method and the result but also on whether well-developed relationships exist between TA and the political and governmental sphere, both on the organizational level and on the institutional level. It is important to strive for such conditions in case of TA-like RRI activities that are sponsored in the context of Horizon 2020.

Understanding interaction mechanisms

Many TA organizations, in particular PTA institutions profile themselves as independent organizations. By taking a relational perspective, our model stresses that creating and maintaining bonds with clients and other relevant actors is crucial for being relevant and having an impact. By acknowledging the dependence of TA on the four social spheres, the way in which interactions between TA and the four social spheres are exactly shaped on the three levels that we distinguished becomes an important research issue. In other words, it is relevant to open up the black box of the interaction between TA and parliament, government, S&T, and society and to study the interaction mechanism used by TA organizations. So the crucial challenge for TA organizations therefore is to deliver independent, trustworthy forms of science-based policy advice and maintain good relationships with the various social spheres

DOI: 10.1057/9781137561725.0010

at the same time. In this way, independent advice, good relationships, and impact on policy can all be achieved in the long run

Lesson 9: Acknowledge the dependence of TA, in order to achieve independent advice with an impact

The challenge for TA organizations is to deliver independent, trustworthy, science-based advice and at the same time establish good relationships with the various social spheres.

Note

1   See also PACITA Deliverable 4.3 'Expanding the TA-landscape' and Chapter 2 of this book.

DOI: 10.1057/9781137561725.0010

OPEN

2

Expanding the TA Landscape – Lessons from Seven European Countries

Leonhard Hennen, Linda Nierling and Judit Mosoni-Fried

Abstract: *This chapter explores socio-political opportunities for and barriers to introducing TA as a support for science and technology (S&T) policy making in seven of the new European member states. Based on interviews with national S&T actors and document studies, the study shows that any attempt to promote and establish TA has to take account of the situations in the countries explored, which differ in many respects from the situation during the 1980–90s when a first wave of TA institutionalization took place at national parliaments in Europe. Elements of 'civic epistemologies' such as a lively public debate on S&T policies are missing in some of the countries explored, and S&T policy making is busy modernizing the R&D system in order to keep up with global competition.*

Klüver, Lars, Rasmus Øjvind Nielsen, and Marie Louise Jørgensen, eds. *Policy-Oriented Technology Assessment Across Europe: Expanding Capacities.* Basingstoke: Palgrave Macmillan, 2016. DOI: 10.1057/9781137561725.0011.

Technology assessment as a means of policy advice is widely established in many Western European countries, whereas in Southern Europe and especially in the new European member states in Central and Eastern Europe, TA structures are often inexistent altogether. The PACITA project, by organizing explorations of existing barriers and opportunities for setting up TA in seven European countries, succeeded in starting up debates about TA among relevant actors and revealed a set of boundary conditions for introducing TA in the national R&I policy-making systems.

The societal situation in the countries explored is different in crucial respects from that of Western Europe during the 1970–80s where (parliamentary) TA institutions were first set up. Thus, not only are elements like a lively public debate on S&T policies missing in some of the countries but also S&T policy makers are busy modernizing the R&I system in order to keep up with global competition.

Our explorations were organized in an 'action research'-like manner – that is, at the same time gathering knowledge about national preconditions for TA while actively intervening by facilitating high-level TA debates or triggering initiatives among relevant national actors. The exploration activities revealed that despite existing barriers, there is a role to play for TA by adapting to and offering support with regard to the existing deficiencies and problems of S&T policy making. Concerns about problems of S&T policy making often result in an explicit demand for 'knowledge-based policy making' in the context of which the concept of TA is welcome as a means to underpin decisions with best available knowledge in an unbiased manner. TA can significantly contribute to ongoing activities of modernizing the R&I system by strategically planning the R&I landscape, evaluating R&I capacities, or supporting the identification of socially sound and robust country-specific innovation pathways. Exactly due to often poorly developed democratic and transparent decision-making structures, TA could find a role as an independent and unbiased player able to induce communication among relevant actors on 'democratic' structures in S&T policy.

To further promote TA, one viable pathway would be continued collaboration – for example, through starting TA projects together with experienced TA countries but also through a continuation of national activities started by the PACITA intervention, such as training practitioners, doing pilot project(s), identifying the specific goals of doing national TA and finding reliable partners in politics but also in other societal spheres (science, industry and civil society).

DOI: 10.1057/9781137561725.0011

Background

Since the 1970s, 'technology assessment (TA)' has been introduced in many Western industrialized countries. Its scientific origins lie in systems analysis and forecasting, but its scope has developed much further – conceptually as well as methodologically (Grunwald, 2009). In those Western European countries that have institutional platforms for TA, the practice of TA is clearly oriented towards policy making, and parliaments are seen as the main client of TA. Motivated by a lack of reliable knowledge and scientific expertise, in many Western countries parliaments have built up dedicated expert units in order to have the capacity to control governments' decisions in S&T policy making. The main impulse for TA in Europe came from the establishment in 1973 of the OTA at the US congress, which mainly carried out expert analysis. After a period of searching for viable European pathways, a range of organizations was founded within European member states from the 1980s and onwards. In contrast to the OTA, some of these organizations focused in part on the involvement of stakeholders and the wider public. (See also the introduction to this volume). Although TA by now is established in many European countries, in other parts of Europe, especially in Southern, Central and Eastern Europe, there are no institutional settings of TA, and also the concept of TA is not used or is even unknown.

One aim of the PACITA project was to explore opportunity structures as well as barriers for TA in countries of Europe without TA infrastructures. To this end, an exploration was carried out in seven European countries (Belgium/Wallonia, Bulgaria, the Czech Republic, Hungary, Ireland, Lithuania and Portugal) to ascertain current needs as well as institutional preconditions for introducing TA in national processes of S&T policy making.[1] The countries explored have very different histories, and in each country debates on TA have very different starting points. In Central and Eastern Europe, TA is established neither in academia nor in policy making. Looking back on the history of Central and Eastern European countries, the differences in Western Europe are obvious. In the planned economy system, the ruling socialist (communist) parties had by far the most significant influence on policy making and in the R&D sector. At best, the Academies of Sciences have been involved in the decision-making process to a modest extent. This involvement was a common feature, although we cannot say that there was a uniform S&T system across these countries. Rather, there were divergent institutional

DOI: 10.1057/9781137561725.0011

systems, especially from the 1980s when cooperation with Western countries became more regular than before enabling relatively open Central and Eastern European countries to introduce new measures – for example, a grant system in research, a dialogue within the scientific community on S&T policy questions and so on. After the transition, the R&D sector and also the Academies of Sciences started to decline due to downsizing of R&D funding and employment. That was followed by a phase of stabilization since the mid 1990s and then by recovery of the R&D sector by the end of the 1990s and early 2000s. As concerns structural changes in the R&D system, a gradual increase in the shares of universities and the business sector can be regarded as the most positive tendency in many Central and Eastern European countries. These stronger R&D actors seem to have a growing role in S&T policy making. However, civil society is only very slightly represented in S&T policy making. On the one hand, this lacking involvement is due to the traditionally peripheral role of the civil society in Eastern Europe, and on the other hand, it is due to the fact that in this region most citizens are more familiar with non-democratic (or 'less democratic') governance systems than with democratic ones.

In the Western European countries of the sample, there are already experiences with 'TA-like activities': In Portugal there has already been some debate on TA in the national parliament as well as in the academic community. While Ireland has a well-developed system of S&T policy advice and consultation, infrastructures explicitly dedicated to TA do not exist. In the Belgian region of Wallonia, there have been debates on parliamentary TA that have been ongoing for many years; however, no institutional setting of TA has resulted so far.

The national studies were conducted from February 2012 to March 2013, and they focused on national political and institutional contexts, existing capacities (actors, organizations and networks), demands and interests in TA-related activities and barriers and opportunities in national/regional contexts. Research methods comprised document analysis, interviews and discussion rounds with relevant actors and stakeholders. The explorations were done jointly by a twin team of researchers from respective national PACITA partners and from an experienced TA partner organization.

It is important to note that the explorations in the countries were conducted from the perspective of different organizations, ranging from Academies of Sciences (Czech Republic and Hungary) to research centres at universities (Ireland, Portugal and Wallonia) and to non-governmental

DOI: 10.1057/9781137561725.0011

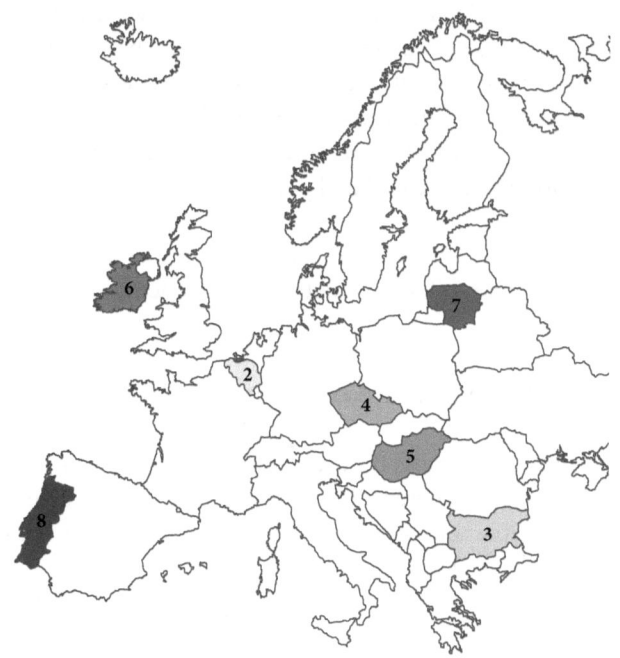

		Population 2011 Mio.	Year of EU entry	Democrat. system since	GDP p.C. 2011 Euro	GERD/ GDP % 2010	Private R&D % 2011	Public R&D % 2010	GBAORD 2010 Euro
1	EU 27	18,61 (tot. 503,7)	–	–	25.100	2,0	61,53	37,51	3.275
2	BE	11,0	1952	1830/1980 (regions)	33.600	1,99	66,3	32,7	2.153
3	BG	7,5	2007	1989	4.8000	0,6	50,0	48,9	96
4	CZ	10,5	2004	1989	14.7000	1,56	62,0	37,4	873
5	HU	10,0	2004	1989	10.100	1,16	59,9	38,4	467
6	IE	4,5	1973	1937	34.900	1,79	68,13	31,9	934*
7	LT	3,24	2004	1990	9.5000	0,79	29,22	70,93	47
8	PT	10,64	1986	1974	16.100	1,59	45,5	44,13	1.763

FIGURE 2.1   *Overview over core economic and R&D data*

Note: * 2007; GERD (Gross Expenditure on Research and Development), GDP and GBAORD (Government Budget Appropriations or Outlays for Research and Development).

Source: ERA watch (http://europa.eu/about-eu/countries/index_en.htm) and Eurostat 2010.

DOI: 10.1057/9781137561725.0011

organizations (Bulgaria and Lithuania).[2] The processes thus had different preconditions in all seven countries. However, the practical aspirations of the project – to spark national discussions on the potential benefits of TA – were successful in all countries insofar as relevant actors were included in the learning process and debates and came to reflect on possible roles for TA in the national policy-making landscape.

The rest of this chapter presents the results of these national exploration processes in a cross-national perspective. The presentation is based on national country reports (for more details, see Hennen and Nierling, 2012).

Societal premises for the setup of TA institutions

Comparing situations across time and space can help to bring attention to those features of the current situation which serve to enable or hinder institutional entrepreneurship. The following comparison between the situation in which Western European countries originally set up TA institutions with the situation today in other European states aims to serve precisely that purpose.

While our comparison of different national settings partly draws on previous analysis of national TA practices (e.g. Delvenne, 2011, Enzing et al., 2012, Ganzevles and van Est, 2012, Vig and Paschen, 2000), the national explorations in the PACITA project had a very practical intent: initiating a debate on TA or even potentially implementing TA in new national contexts. For this purpose, the most important background information is the societal situation in the 1970s and 1980s which led to the establishment of a number TA institutions in the US and in Europe. This is the historical situation to which we compare the current situation in the countries that we studied.

We consider the following societal features of Western Europe in the 1970–80s to be relevant reference points for current discussions on institutionalizing TA capacities:

1   *Highly developed and differentiated R&I systems existed, which had strong backing from governments aiming to strengthen the international competitiveness of their national economies.*
2   This was reflected in the setup of research ministries, the growing public funding for R&I and the increasing importance of R&I in parliamentary standing committees.

DOI: 10.1057/9781137561725.0011

3 *A strong and critical interest of the public towards S&T issues was prevalent.*

4 Not only was this critique articulated on the general level, but also citizens' initiatives on different political levels (local-national) fought for participation in planning decisions as well as S&T politics because they were considered to interfere with citizen's rights.

5 *Interdisciplinary, problem-oriented science gained influence in several academic fields.*

6 The term 'sustainable development' served as a key term for this kind of 'new' research.

7 This development in academia also led to academic support for 'TA-like "hybrid science" and policy-oriented research' (Hennen and Nierling, 2014b).

Within this societal situation arose a strong demand by policy makers for reliable knowledge on scientific and technological developments, as well as for methods to cope with public concerns.

In some countries, these demands led to the establishment of institutions which supported national parliaments with non-partisan scientific advice. In other countries, they led to institutions organizing and raising public debate. Thus, TA bodies where institutionalized in different ways each relating to national parliaments and governments (again, see also Chapter 1).

Against this background, the results of the comparative study will be presented below with the aim of showing differences and similarities among the countries with regard to the reference points identified above. First, the current R&I landscape and national R&I performance including ongoing strategies of modernizing and restructuring the R&I systems as well as problems and deficits of the current systems will be described. Second, the levels and central features of political and public debate on S&T will be highlighted. Finally, already existing structures of TA-like research and/or policy advice will be presented.

National R&I landscapes: R&I performance, modernizing strategies and deficits of the current system

R&I performance

In all the countries that we analysed, R&I topics are generally high on the political agenda, reflecting the importance of R&I for economic

DOI: 10.1057/9781137561725.0011

development and its relevance for catching up with increased global competition. However, the broader S&T policies are developed in a difficult situation. On the one hand, in most of the countries involved, the economic situation is difficult. With the exception of Ireland and Wallonia, all national economies are lagging behind the EU28 average development in terms of their gross domestic product (GDP). Furthermore, due in part to their relatively weak economic performance, the expenditures and investments in R&I of these countries are (in some cases significantly) below the European average. For the Central and Eastern European countries, this is undoubtedly due to the fact that their economic modernization is a disappointingly slow and conflicting process, involving political and social tensions. Thus, economic growth in these countries seems to be rather fragile, economic forecasts. The people in these countries are disappointed by this development because people had expected fast-paced improvements in their quality of life. Instead, citizens still experience many constraints in different fields: political (democracy-deficit), social (poverty, problems in health care, education, housing and so on) and human-economic factors (high proportion of unskilled workers, lack of job prospects and permanent gap between the developed and backward regions). However, some countries, such as the Czech Republic and Hungary, have already achieved considerable progress in increasing their share of private R&I investment. Both Portugal and Ireland are in a process of restructuring their economies from a model dominated by agricultural structures to a modern knowledge-based economy – and Ireland has been extremely successful in this respect in the last two decades. However, precisely because they were in the middle of a complex and expensive process of restructuring, the financial crisis struck these countries hard and the strain on public budgets led to a decrease in R&I expenditures. Belgium (Wallonia) is the only studied country that can be regarded as being in a position similar to the average European capitalist economies, especially because Wallonia is undergoing a shift from traditional industrial structures to an S&T-based economy and invests heavily in research clusters in order to manage this transition.

Modernizing strategies

Generally, building up the economy sets the main frame for R&I policy making. All the countries that were explored have set up national innovation strategies to modernize the R&I system, attract private investments

DOI: 10.1057/9781137561725.0011

and improve competitiveness. The key targets listed in governmental R&I programmes and strategies can also be read as a list of the typical deficiencies of R&I governance, infrastructures and strategies.

In most of the countries that were explored, a set of institutions exists, which give advice to the political sphere (policy makers and government) on a regular basis, be they specialized expert committees connected to ministries, specific funding programmes or national science policy councils. National R&I councils mainly represent Academies of Sciences, industry, universities, public administration and the non-profit sector. They have been established to coordinate reform strategies and to advise the government. In the case of the Czech Republic, the Council for Research, Development and Innovation has almost taken over the role played by a ministry and is more or less designed to centralize the system of R&I and even to take over micromanagement tasks (Pokorny et al., 2012: 69). Because research councils mainly represent academia, industry and public administration, they can be regarded as an element of academic self-administration and expert policy advice. The involvement of industry is meant to establish closer relations between public and private research bodies in order to improve innovation performance. Advice is mainly addressed to the government and rarely to the national parliament.

It is apparent that strategic advice with regard to the future development of research and innovation strategies given by these institutions is motivated by national efforts to improve the competitiveness of the national economy ('economy first'). Compared to these activities, policy advice with regard to future (controversial) technological or scientific development is of minor relevance. This is in line with the fact that foresight methods are frequently applied by governmental agencies to assess the economic strategic planning (for instance, the recently published 'National Research Infrastructure Survey and Roadmap' in Hungary), whereas TA as a means of policy advice is almost unknown in many countries.

Problems and deficits of current R&I governance systems

The country studies reveal a plethora of activities to modernize R&I structures as well as R&I governance systems. The problem is often not a lack of institutional reforms and new agencies but rather a lack of functionality and efficacy. Interviews and workshops revealed scepticism with regard to the effectiveness of newly established systems and strategies by actors from academia and policy making, as well as industry and civil society.

DOI: 10.1057/9781137561725.0011

In general, the effectiveness of strategies seems to be compromised by discontinuity and a lack of focus mainly because of quickly changing political agendas driven by short-term tactics and by quickly shifting political power. Discontinuity in setting up reforms is reported as being explicitly a main weakness of R&I policies for Hungary, Bulgaria and Lithuania, due to shifting parliamentary majorities or a general lack of coordination strategies. Thus, innovation strategies are often perceived as 'activism' since they apparently result in constant reorganization of strategic planning. For example, each government in Hungary initiated a reorganization of the policy making and advice structure in R&I at least once in their four-year term (Mosoni-Fried et al., 2012: 113).

Deficiencies in existing advisory systems

A lack of transparency in decision-making processes, and thus of public trust in and legitimacy of policy making, is reported in all countries. A strong need to improve the current situation of national policy advice is expressed in the Bulgarian and Portuguese reports with regard to the legitimacy and transparency of political decisions, as well as setting up missing communication channels between science, politics and the public. In most of the countries that were studied – for instance, Bulgaria – S&T expertise is typically provided internally by governmental staff at the respective ministries. On rare occasions, external expertise is asked for on an ad hoc basis, and even in these cases, the process remains opaque to the wider public (Kozarev, 2012: 42). Although a number of institutions often provide policy advice (for example, a formal advisory body of the government or other national councils) and although an occasional demand for scientific advice from the political sphere exists (for instance the government or parliamentary commissions), there seems to be no institutionalized or 'routinized' ways for constant policy advice. Rather, communication channels among scientists, policy makers and other potential knowledge providers are characterized as 'fragile and dependent on the continuous will of interacting between specific stakeholders' (Almeida, 2012: 235).

Even if processes are formally transparent, with relevant documents for decisions being publicly available and consultation with experts taking place, many interview partners experienced a lack of accountability. It appears that administrations act without taking the arguments

DOI: 10.1057/9781137561725.0011

of consultations (be they expert or public) into account. A certain level of distrust in governmental performance on the part of academics or other experts appears to be significant in many of the countries that were explored. In Central and Eastern European countries, this may be related to a great extent to the conflicting character of the ongoing and long-lasting political transition period from a non-democratic system to a democratic one. In Ireland, the reported lack of transparency and public involvement in R&I policy making may rather be rooted in a lack of cooperative traditions and the remaining authoritarian political culture clashing with the country's rather new and fast emergence as an R&I economy. Thus, apparently, the highly developed Irish system of advisory bodies and agencies has not yet opened up to the wider public and remains a closed deliberative circle of the executive branches of government and related expert communities.

Public debate on S&T

Complaints about a low level of political as well as public debate on S&T issues are widespread in interviews and workshops. Generally, a 'systematic integration' of S&T issues in a societal discourse that includes all relevant groups (politicians, scientists and society) seems to be missing. Conflicting factors very well known from Western democracies, such as long-term S&T issues versus short-term political agendas, may have an even stronger influence in countries where democratic structures and cultures are still in transition. Other factors mentioned are clearly connected to the communist heritage in Eastern and Central European countries, such as a 'lack of a debate culture and debate traditions' (Kozarev, 2012: 37) (Bulgaria), or a general scepticism with regard to public debate rooted in the national political culture (Lithuania). Platforms for controversial debate on S&T issues (also in parliament) are missing, and the lack of transparency in decision-making structures – mentioned above – clearly leads to a restriction of debate to a closed circle of experts. The conditions for public debate on S&T are more favourable in Ireland and Wallonia. In Ireland, the interest of politicians in citizen participation has grown remarkably in recent years (O'Reilly and Adam, 2012: 159) due to current technological conflicts at the local and regional levels. In the ongoing political discussion about setting up a TA institution in Wallonia, public involvement is a central topic for those policy makers who are involved.

DOI: 10.1057/9781137561725.0011

It adds to the notion of a lack of public debate that public interest in S&T issues is reported to be low in most of the countries. The latter notion is sometimes coupled with a well-known prejudice against laypeople who are regarded by policy makers as being 'emotional and incompetent' (Mosoni-Fried et al., 2012: 126). The notion of a relatively low interest in S&T is supported by European survey data (TNS Opinion & Social, 2010, 2013): the citizens of the countries that were analysed here are less interested in S&T issues than is the average European: they less often read articles on science in newspapers, in magazines or on the Internet, with only Belgium and Ireland being above the European average (TNS Opinion & Social, 2005: 23, 2013: 6). Moreover, for a broad majority of respondents from the countries that we studied, the involvement of experts (scientists, engineers and politicians) is regarded as the most appropriate way to make political decisions in S&T.

The reported 'lack of debate' is to some extent modified by the fact that the country studies outline a broad range of contested S&T issues, such as genetically modified organisms (GMOs), energy policy, waste management and food safety. Specific implications of technologies such as information and communication technologies (ICTs) or ethical concerns in controversial fields such as assisted reproduction were also debated within national contexts. Furthermore, locally or regionally embedded large-scale technological projects such as a dam or an oil pipeline became a subject of national debate. With regard to the development of citizen participation, it should be noted that there are different historical contexts in Western Europe as opposed to the post-communist countries (see Hennen and Nierling, 2014b).

Existing structures of TA-related research and policy advice

The scientific landscape in all post-communist countries in our sample is still very much influenced by the prominent role of the national Academies of Sciences. Although none of the academies were active in the field of TA prior to the PACITA interventions, at least in the Czech Republic and in Hungary, there are traditions of problem-oriented and interdisciplinary research, as well as of applying methodologies relevant to TA (foresight, future scenarios, indicators for sustainable development and more) at the national academies and universities. Since 1998, Hungary has had a strong foresight tradition (Mosoni-Fried et al., 2012: 116), and the work of the academy has taken up current societal topics in

DOI: 10.1057/9781137561725.0011

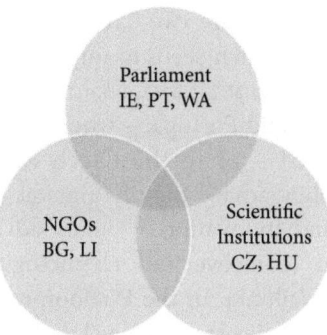

FIGURE 2.2   *Possible pathways towards TA*

the Hungarian context, such as waste management, food safety, climate change and the red sludge catastrophe in 2010. In the Czech Republic, some institutions already have more concrete experience with TA and TA-like activities, such as the participation of the Czech Academy of Sciences in EU-funded projects on TA, the establishment of the Czech Council on Health Technology Assessment at the Ministry of Health, as well as the Czech participation in various European foresight activities.

In Lithuania and Bulgaria, the science academies currently seem to have a less influential role and also less experience with interdisciplinary and problem-oriented research. In Lithuania, the roles of the Academy of Sciences and of the research council seem to be more formal. Policy advice is provided to the parliament as well as to ministries. However, for the academy, it is more important to take up the mission to promote science and scientific literacy in the wider public (Leichteris and Stumbryte, 2012: 195). In Bulgaria, the Academy of Sciences currently faces major internal restructuring combined with severe problems in scientific knowledge production, which led to the low public reputation of scientists and also to an erosion of trust in scientific institutions in recent years (Kozarev, 2012: 43).

In contrast to the Central and Eastern European countries, in Ireland and Wallonia there are quite a few scientists active in TA-like approaches, such as problem-oriented applied research in the fields of science in society, STS studies, or environmental studies – including a set of PhD programmes, as well as a range of research institutes working in this field. Similarly in Portugal, the most active institutions in fields related to TA are academic ones. Portugal thus has an international PhD program in the field of social sciences and technologies that focuses

DOI: 10.1057/9781137561725.0011

specifically on TA, and there are two TA-related stakeholder networks (GrEAT[3] and Bioscience) which seem to imply a strong academic focus on TA in Portugal (Almeida, 2012: 235f, Moniz and Grunwald, 2009).

In contrast to Bulgaria and Portugal – where improved organizational procedures are requested – or to the Czech Republic, Hungary and Lithuania – where policy advice mainly aims at strategic planning of science, technology and innovation – policy advice dedicated to the assessment of certain (controversial) technologies is already established in Ireland and Wallonia. In the Walloon region, a wide range of governmental advisory bodies are active with regard to S&T in different fields for 'technology guidance' or in the field of environmental assessment. However, the level of cooperation between the different entities appears to be quite low, and their focus is quite specialized. For Ireland, it is reported that since the mid 2000s, S&T policies have increasingly been questioned, which also implies an increased interest in 'strategic intelligence tools', including TA and foresight (O'Reilly and Adam, 2012: 160). More recently, the wish for public involvement was renewed during public upheavals due to the protests against shale gas exploitation in 2012. In this context, policy makers started initiatives to enforce public involvement to learn about the motivation of local protests and citizens' demands (O'Reilly and Adam, 2012: 160).

The deficit in terms of societal involvement in R&I policy making is aptly reflected in the fact that the role of parliaments in R&I policy making is reported to be quite low in most of the countries that we explored. In most of the countries, the focus of parliamentary committees that are in charge of R&I policy making is mainly on higher education. Parliaments are also reported not to have the resources to support their debates with the necessary knowledge on R&I issues. In most cases, parliamentary committees only occasionally organize hearings to improve the knowledge base for debates. Connected with the weak role of the parliaments is apparently also a lack of permanent structures at the interface between science, society and policy making, as reported for Portugal (Almeida, 2012: 230). It is difficult to draw conclusions from the country studies regarding the reasons for the low involvement of parliaments. Explanations given in interviews, such as MPs' lacking a personal background in S&T, appear to be inadequate. Instead, we might speculate that the low level of public engagement in R&I issues, combined with the general consensus in which R&I is seen as the best guarantee for national economic development, together have the effect

DOI: 10.1057/9781137561725.0011

of preventing interest in a thorough deliberation on risks and benefits from arising. This lack of interest might then in turn explain the lack of parliamentary debates.

Ways forward: Possibility structures for TA

For the Central and Eastern European countries, it can be stated – albeit with a few notable exceptions, such as the Czech Republic (see above) – that the concept of TA was widely unknown before the PACITA project introduced it. An aim of our exploration was to first make the relevant actors aware of the idea behind the concept of TA and its practical workings as a tool of policy advice in order to encourage them to reflect and discuss the possible relevance of the concept in their national academic and policy making setting as a second step. This was done with quite some success at the national workshops that were organized as part of the exploratory research. The discussion of the TA concept and its societal outcomes and benefits was continued in the course of the PACITA project, namely by a parliamentary hearing on a European Future Panel on Public Health Genomics as well as by a stakeholder process on urgent questions of the Ageing Society (see Chapter 6 and Chapter 7). Whereas the topics provoked different responses dependent on national political agendas, the format of public dialogue raised intense interest in participatory TA methods in all countries, which resulted, for example, in broad media coverage of the TA events in Hungary and in a stronger commitment of the Hungarian Academy of Sciences to the idea of TA.

Possible institutional models

When it comes to policy options, especially with regard to the further development of a TA infrastructure, the country studies propose different paths which are categorized in the following sections.

Supporters of parliament (Ireland, Portugal and Wallonia)

In Wallonia, Ireland and Portugal, members of parliament or of parliamentary committees expressed their interest in TA, thus parliament was selected as main addressee for TA activities in these countries. The process is furthest advanced in Wallonia where a parliamentary mandate for TA was given in 2008. Ireland and Portugal are at the beginning of

DOI: 10.1057/9781137561725.0011

such a process, as both parliaments have expressed an interest in TA. In both countries, the parliaments have a rather weak political role. Whereas in Ireland TA is regarded as a possibility to strengthen the role of parliament (O'Reilly and Adam, 2012: 162), in Portugal the advantages of a TA unit in parliament is seen as a possibility to support the country's 'political, social and economic' development (Almeida, 2012: 237).

In all three countries, the explorations advise using existing institutions for future TA activities to draw on national academic expertise in S&T. Furthermore, a special interest is expressed for participatory aspects in a future TA unit, either to create the first, to improve national experience with methods of participation, or to include relevant stakeholders and the public in political decision making in S&T in the future.

The innovative explorers (Bulgaria and Lithuania)

The national recommendations developed for Bulgaria and Lithuania present a new model for a national TA landscape: the network model. The model basically implies that a network of existing institutions collectively take on the task of delivering TA services coordinated by one organization perceived as legitimate by all involved. In both countries, there was very little prior experience with TA or TA-like activities. However, during the research activities, TA was identified as 'an unrecognized need' (Leichteris and Stumbryte, 2012: 200) by some of the relevant decision makers. The main function of such a network model is to raise awareness of S&T topics in the public and by decision makers in relevant political fields. Both countries consider it helpful to start with a pilot project (similar to the starting phase of some established TA institutions in the 1980s and 1990s; cf. Ganzevles and van Est, 2012) in order to 'prove' the national relevance and to increase the understanding of the TA concept and its 'products'. In Lithuania, this 'proof of concept' is currently set into practice by a group of institutions form academia, public administration and civil society with a range of policy briefs prepared for policy makers to 'showcase' the use of TA (see also Chapter 3).

The institutional traditionalists (Czech Republic and Hungary)

The Czech Republic and Hungary make up a third group. In both countries, the Academies of Sciences are decisive players in the field of S&T policy. Furthermore, the national academies in both countries have been in contact with TA or TA-like activities (especially foresight and S&T

DOI: 10.1057/9781137561725.0011

studies). Both evaluate the 'system barriers' (Pokorny et al., 2012: 80) in the current political context as being quite strong and are thus pessimistic about the future establishment of a TA unit. Barriers to be dealt with include a lack of options for national funding, a lack of trained personnel, but also a general lack of interest from the decision-making sector in S&T as well as the public. Interestingly, during the course of the PACITA project, triggered by accompanying activities such as practitioner meetings and participatory events, the academies in both countries got more and more convinced and thus interested in TA-like activities (see also Chapter 3).

Future perspectives for national TA capacities across Europe

Looking back in history, it becomes clear that TA must be understood as a reaction to the failure of a 'technocratic' concept of the relationship between science and politics dominant in the 1950s and 1960s, which relied on scientific knowledge as a safe and sufficient ground for 'rational' policy making. Thus TA, as it were, has always been taking into account the inborn uncertainty and underdetermined character of scientific knowledge with regard to complex practical (political) problems as well as the indispensable need to take into account different (and often conflicting) values, normative claims and expectations held by societal groups. The transparency of the TA process and openness towards the public, involving a broad scope of interests and values have been essential features of the TA concept right from its start.

Our country studies give quite clear indications that the context for TA initiatives (not to speak of processes of institutionalization) is in many respects different from the conditions that were prevailing when the first wave of TA institutionalization took off. In most of the countries that we explored, the concern is not about the further development of an already strong R&I system as it was in Western Europe when TA was established. It is rather about building new structures or about fundamentally reorganizing existing structures in R&I. In Eastern and Central Europe, the R&I landscape is in transition (as it is for other reasons in Ireland and Portugal), and it is less about 'protecting' societal needs and values against the dynamics of S&T. Instead, what is in focus is instigating dynamics and exploring innovation paths to keep up with globalization

DOI: 10.1057/9781137561725.0011

pressures and to generate economic growth. The social impact of S&T comes into perspective less in terms of environmental or health risks and ethical issues and more in terms of supporting societal welfare. Thus, TA is expected to provide support with strategic thinking on robust R&I structures, options for innovation policies and the evaluation of existing structures and practices. It is not by accident that whereas TA often is not very well known in the countries that we explored, 'foresight activities' have been widely promoted in some of them.

With the exception of Wallonia and Portugal, parliaments are not active in taking up TA as a means to strengthen their own role. In the beginning of the PACITA process, parliaments were often also not regarded by TA-interested actors as appropriate places for TA activities. This attitude has changed a bit in the course of the project. By now, all partners have increased the cooperation with national parliaments and established connections with national parliamentarians that support the vision of national TA capacities. Countries without established TA institutions have drawn the lesson from the practice of PTA countries as well as from the history of institutionalization of TA all over the world (Hennen and Nierling, 2015), namely that acceptance, acknowledgement and support of TA demand high quality TA activities, on the one hand, and distinguished individuals, mainly politicians who are interested in independent policy advice on technology issues, on the other. There are not too many potential political TA partners in the countries that we have explored so far, but already a few of them are able to do a lot.

Throughout our country studies, a lack of democratic structures in S&T policies is often perceived as well as a lack of communication and cooperation among relevant actors (academia, government, parliament and civil society organizations (CSOs)). TA then comes into perspective as a means of unbiased information of discourses (such as knowledge-based policy making or responsible innovation) or a platform to establish a democratic (public) S&T discourse (independent of reflections on its institutional setting).

In contrast with the conditions under which TA began, S&T is far less an issue of lively public discourse and activism. Whereas the present relatively low public engagement in S&T debates in Western Europe comes with an established system of professional and public authority bodies dealing with risk assessment and ethical issues, such structures are missing in the countries explored here (with the exception of Wallonia). For those examples of public controversies that were reported, it is on the

DOI: 10.1057/9781137561725.0011

one hand often stated that they are characterized by a lack of platforms for constructive interchange of actors including CSOs and laypeople. TA is expected to play a role in this respect. On the other hand, 'the public' often comes into focus with complaints about a lack of interest in, and knowledge about, S&T issues. As much as this might be in line with a well-known attitude of scientific elites and the prevalence of the so-called deficit model of public understanding of science, this might also indicate a specific problem connected with a lack of trust in democratic structures and with a distance to the political process that goes beyond the usual disenchantment with politics. In all the countries that we explored, there is, to various degrees, a lack of tradition in public debates on S&T as well as a relative lack of structural channels or platforms for public debate (including media and CSOs). Thus, 'stimulating public debate' as a mission of TA may gain particular importance here.

On the practical political implications of these features of a – so to speak – new 'TA habitat in the making', we see the following challenges in terms of practical expectations that TA has to react to:

▶ *Ongoing, often not well-coordinated activities of governments to build up or restructure the R&I system*: In this respect, TA is often explicitly expected to contribute to strategic planning of the R&I landscape and the evaluation of R&I capacities.

▶ *Innovation policies to improve competitiveness in the context of globalization and crisis ('economy first')*: TA would have to position itself with respect to these activities by providing support for identifying socially sound and robust country-specific innovation pathways ('constructive TA') and contribute to lower costs of trial-and-error learning.

▶ *Poorly developed democratic and transparent decision-making structures*: TA could find a role here as an independent and unbiased player able to induce communication on 'democratic' structures in S&T policy among relevant actors.

▶ *The challenge of 'involving the public'*: In this respect, the motives of democratizing policy making are often merged with 'paternalistic' motives of 'educating the public' (media and laypeople). The latter nevertheless may indicate a real problem of broad public unawareness regarding the democratic relevance of S&T politics and the extent to which TA's mission of 'stimulating public debate' can adapt to that problem (without becoming 'persuasive').

DOI: 10.1057/9781137561725.0011

▶ *In transparent decision making, lack of trust in democratic structures, lack of competences and bounded rationalities of relevant actors, lack of strategic long-term thinking*: All this results in an explicit demand for 'knowledge-based policy making' in the context of which the (not very well-known) concept of TA is welcome as a means to underpin decisions with the best available knowledge in an unbiased manner. Specific ideas about how to institutionally build it into the existing system are, however, missing, and it might well be that in terms of institutional solutions none of the models so far realized in Europe might be appropriate.

In general, TA has to be responsive to the given policy context and the expectations and demands expressed in the countries that we explored. However, 'being responsive' to national expectations should not imply giving up a certain (normative) core of TA as a concept. TA risks becoming an 'empty signifier' if its proponents seek to respond to any and all demands for 'rational' decision making and planning expressed by policy-making bodies and authorities. TA as a concept implies the role of a critical observer of R&I policy-making activities, which necessarily asks for some institutional independence in order to provide space for reflection beyond short-sighted political agendas and openness to a broad spectrum of perspectives being applied in assessment processes.

Notes

1   For more details, see L. Hennen and L. Nierling (2012).
2   The evaluation is given from a specific organizational perspective and does not claim to fully reflect national debates or newly evolved initiatives.
3   Grupo de Estudos em Avalicão de Tecnolgia (GrEAT) is a Portuguese network on TA (see http://avaliacaotecnologia.wordpress.com/).

3

Adopting TA in Central and Eastern Europe – An Organizational Perspective

Lenka Hebáková, Edgaras Leichteris, Katalin Fodor and Ventseslav Kozarev

Abstract: *Hebáková et al. provide from an insiders' perspective the process of adopting and adapting technology assessment to the practices of an already existing institutions. The strategic thinking of four very different organizations in four Central and Eastern European countries is candidly described and contrasted. The authors challenge the notion of technology assessment as a set of ideas and practices to be adopted en bloc. Rather, TA provides a package of inspiration that may help organizations to broaden their missions within the field of national science, technology and innovation policy to include, for instance, parliamentary policy support, facilitation of stakeholder dialogues or citizens' participation.*

Klüver, Lars, Rasmus Øjvind Nielsen, and Marie Louise Jørgensen, eds. *Policy-Oriented Technology Assessment Across Europe: Expanding Capacities.* Basingstoke: Palgrave Macmillan, 2016. DOI: 10.1057/9781137561725.0012.

International communication among circles of professionals in policy and administration has always been a core impulse for the development of new institutional forms. But an equally universal prerequisite for the adoption of such new forms is the successful adaption of these forms to the national context. In this process of adaption and translation, entrepreneurs within existing institutions play a crucial role. It is their commitment and energy that propel institutional reforms, and it is their creative negotiation of the 'space of opportunity' which helps to shape nationally acceptable solutions for adoption of new institutional forms.

In this chapter, we zoom in on the process of attempting to adapt TA to the institutional realities of the Central and Eastern European partner countries. We recount this process such as it was experienced by the PACITA partner organizations from those countries.

The inside scoop: taking TA on board in existing organizations

For the Technology Centre of the Academy of Sciences of the Czech Republic (hereafter, 'the Centre'), established 1994 as part of early reforms in the post-soviet era, taking part in PACITA has created a lot of internal interest and debate concerning the concepts and practices of TA. But far from being seen as any revolutionary change, TA is seen to fit naturally alongside already existing organizational priorities. To explain this fit, it is useful to understand the role of the Centre. The Centre is a key organizational player in the development of the Czech STI governance system that provides analytical support for several governmental actors in that field. The Centre often acts as an intermediator among different government bodies involved in STI policy formation, and it serves as a connector to international STI collaboration, serving for instance as the National Contact Points Centre for European Framework programmes for research and providing support to analyses of international innovation systems conducted by EU institutions, UNIDO, OECD and so on. In terms of practices, the Strategic Studies Department, which was directly involved in PACITA, has long provided key services, such as policy analysis and evaluation, bibliometrics and foresight studies. It was the Centre's experience with foresight and its international networks with practitioners from other countries which provided the basis for

DOI: 10.1057/9781137561725.0012

the Centre's entrance into the PACITA partnership. And it is alongside this base of experience that the concepts and practices of TA are now becoming part of the organizational priorities of the Centre. From the point of view of the Centre, TA and foresight methodology are seen as part of a continuum of similar activities where the contribution of the TA tradition is its stressing the societal dimension of foresight, the value of participation and the idea of including parliament more directly in the policy process concerning STI issues. The various PACITA activities, including the example projects (described in part II of this book), have provided a welcome opportunity to seek out contacts with parliamentarians. Parliamentary debates concerning TA that were facilitated by the Centre have started a longer-term discussion about possible ways of including TA in the EU Operational Programmes funding research, development and education, as well as the possible role of the Centre as a support function for parliament. But again, this should be seen as a natural expansion of the already crosscutting institutional role of the Centre in the national STI policy system.

In Hungary, the participating Hungarian Academy of Sciences (HAS) has an institutional history that predates the former communist system by a hundred years. As such, it is viewed by the majority of stakeholders in the STI field as well as by the citizens as the most highly trusted public institution. This means that adopting TA takes place on a basis of an already well-established institutional platform and a highly vibrant range of international connections. Because the Academy is already a research-performing organization of significant size, which already has scientific policy advice role on the national level and international cooperation as a core part of its mission, TA is seen perhaps more as an addition to its internal palette of activities and competences than as any significant change in its role vis-à-vis other societal and governmental stakeholders. The Academy's culture is one of strong traditions and a high regard for the role of the scientific expert. The most salient feature of TA for the Academy has therefore been the overall idea of increasing the transparency of STI decision making and offering a platform for dialogue on socially relevant STI-related issues. Participating in PACITA has occasioned reflections on the usefulness of opening up to societal stakeholders in order to increase the societal responsibility of STI policies. Taking up relations with parliamentary representatives proved to be a fruitless effort during the PACITA project. It was partially due to the engagement of the potential partners in the period of the parliamentary

DOI: 10.1057/9781137561725.0012

elections in 2014. Here, the organization experienced that debating the concept of TA in broader terms was not as useful as were the example projects, which illustrated much more clearly the value of doing TA. Because the Academy is connected with the capacity-building effect of doing TA events, it will prioritize the creation of further concrete projects to serve as examples and to strengthen the human resource build-up internally in the organization. Such concrete projects, moreover, also serve to build networks of people interested in the specific policy issue being treated. This TA networking function is a key add-on for a few Hungarian institutions and, as such, is a valued outcome of the project for the Academy.

In Lithuania, in contrast to the well-established Czech and Hungarian partner organizations, the Knowledge Economy Forum is a relatively newer organization. The Forum plays an ever-changing role of pushing the development of the national STI institutions, a role which was first defined at the Forum's establishment a little more than a decade ago in terms of promoting business interests. With increasing funding going to early-stage R&D in support of innovation, this early mission was in some sense accomplished, and new steps towards further advancement of the national innovation system had to be found. In this situation, the opportunity presented by PACITA of considering in depth the role that TA may play in the institutional development of the country was well timed. Compared to the 'first wave' of TA institutionalization in Western Europe, the Forum's origins as an interest organization might have been thought to preclude adoption of the traditional role of a TA organization, where 'neutrality' has been seen as a central virtue. But from a reformist perspective, it makes sense in the Lithuanian context to promote greater institutional and political attention around societal issues related to STI. Authors on national systems of innovation have long stressed the need to build trust through cross-institutional dialogue. And social and environmental issues become increasingly important dimensions of international product competition. The Forum has thus come to see it role as promoting in a more complex manner the interests of its constituents through the development of dialogical forms of policy formation that take into account environmental and social issues related to the innovation-driven economy. In promoting this new focus, the Forum has developed a 'network model' for TA (see Chapter 2) in which the plural landscape of many small institutions engaged in STI policy are drawn together around the formulation of

DOI: 10.1057/9781137561725.0012

policy recommendations for how to take into account broader impact dimensions of policy. Establishing the legitimacy of this solution is an ongoing process in which a balance is to be found with the institutions that remain from the communist era.

In Bulgaria, the Applied Research and Communications Fund (ARC Fund) has established itself as the premier research organisation into issues related to science and innovation policy. It was founded in 1991, and it is among the first post-communities-independent non-governmental organisation in Bulgaria, as well as one of the very few still actively in operation. Since its inception, its ambition has been to support the development of the knowledge economy in Bulgaria and in Europe by introducing new policy concepts and innovative policy-making tools (such as foresight) by promoting policy consensus among actors in government, industry and civil society and by helping build the capacity of various professional groups. PACITA-project objectives were highly in line with these ambitions, and being a partner in PACITA further enabled ARC Fund to extend its methodological capacity by focusing more closely on the interlinkages among policy, science and technology, especially by stimulating civil society input through various participatory engagement methods. Although the concept and significance of technology assessment have gained in popularity, technology assessment as such is still not widely recognisable among stakeholders. Particularly in parliament, assessments of specific technologies have been performed with regard to social impacts. However, the scope and depth of this analysis were relatively narrowly defined and confined to a specific political agenda.

There still exists the need to define properly the best 'client' for technology assessments as parliament alone is often only the last among a range of policy actors who promote a specific policy development. This is in large part due to the structure of the legislative decision-making system, which facilitates much of the expert-based work to be done within ministries and other government agencies before it is submitted as a proposal to parliament and then debated and enacted within a relative short time frame. This presents ARC Fund with the opportunity (and challenge) to interact with a number of policy actors and to perform a number of functions, functions including expert identification and networking, quality assurance, (science) communication and policy uptake promotion, in addition to organisational and analytical tasks.

DOI: 10.1057/9781137561725.0012

Possible new approaches to the adoption of TA

A short opinion poll was taken at the end of the PACITA project among the countries, and these have been classified according to their self-evaluation of the institutional positioning in the STI policy advice. The opinion poll was based on four categories, defined as follows:

▶ *Content marketer* shall give politicians their desired 'shortcut', but the content marketer institution shall make it as methodologically correct and objective as possible within the limits of available financial and human resources.

▶ *Eyes opener* shall give politicians a glimpse what is going on at EU level or in other European countries and raise awareness on important issues. TA can be understood as a broad set of practices aimed at informing, shaping and prioritizing technology policies and innovation strategies, by deliberately appraising in advance their wider social, environmental and economic implications.

▶ *Lobby organization* shall aim at building up big coalitions and putting issues on political agendas, not at defending particular interests. Networking shall be used intensively to make personal relationships with policy makers and to form some general positive public opinion on knowledge-based policy making. If the resources allow, policy evaluations can be performed – showing shortcomings of current policies and providing general recommendations for action.

▶ *Knowledge sharer* shall concentrate on cross-border European exchange. There will always be a constant need for various examples of how one or another issue is solved in other countries. If Germany, Austria, The Netherlands or some other TA country can afford large-scale research on the impact of technologies developed in their countries on society in general – in the case of Eastern European countries and their budgetary constraints and undeveloped R&D systems – then adapting already existing EU knowledge into the local context might be a more feasible solution. That's why cross-European cooperation of TA-like institutions is so important.

Representatives were asked to prioritize what is the likelihood that their institution would take over a particular function in the near future. The results are presented in Table 3.1 below.

DOI: 10.1057/9781137561725.0012

TABLE 3.1   *Likelihood of institution taking over a particular function*

Function/Country	Hungary	Czech Republic	Lithuania	Bulgaria
TA as a 'content marketer'	4	1	1	3
TA as an 'eyes opener'	2	4	4	1
TA as an 'lobby organization'	1	2	2	horizontal
TA as a 'knowledge sharer'	3	3	3	2

By way of concluding this inside look, it is clear that adopting a TA role does not equate to taking a step up an evolutionary ladder. Rather, the tradition of parliamentary TA provides ideas and practices, which each organization cherry-picks from in ways that suit their organizational style and institutional role. From the point of view of these organizations, the ambition to expand TA across Europe thus provides a welcome source of new inspiration for already ongoing processes of institutional development and refinement in the STI field.

DOI: 10.1057/9781137561725.0012

OPEN

4
Technology Assessment for Parliaments – Towards Reflexive Governance of Innovation

Danielle Bütschi and Mara Almeida

▶

Abstract: *Bütschi and Almeida explore TA's importance for policy making today, taking into consideration parliamentarians' needs and expectations. The chapter highlights the challenges policy makers have to face when dealing with science, technology and innovation and discuss how TA can address them at an institutional level. These challenges go beyond the complexity of STI policy issues. Globalization challenges policy making on science and innovation as issues spill over national boundaries. As innovation is increasingly expected to foster growth and employment, policy making has to foster innovation and mitigate risks. And last but not least, the financial crisis is challenging parliamentary democracy with top-down fiscal crisis policies. This is where the advanced dialogical and transdisciplinary practices of TA may add value that other advisory practices cannot.*

Klüver, Lars, Rasmus Øjvind Nielsen, and Marie Louise Jørgensen, eds. *Policy-Oriented Technology Assessment Across Europe: Expanding Capacities*. Basingstoke: Palgrave Macmillan, 2016. DOI: 10.1057/9781137561725.0013.

DOI: 10.1057/9781137561725.0013

Science, technology and innovation play an increasingly important role in national and European political agendas. In times of economic and financial crisis, policies in support of research and innovation are being considered as key elements for economic growth and competitiveness, supporting the prominence of innovation in the policy agenda of many countries and of the European Union. At the same time, science and technology developments are challenging existing public policies and legislation due to the impact that they may have in terms of environmental sustainability or social equality. For instance, advances in biomedicine and information technology are leading to ambitious and powerful innovations which will affect health-care systems in Europe. Surveillance technologies used to increase national security may pose problems in terms of data protection and privacy.

The expanding role of science and technology in policy making challenges the role of parliaments in democracy. It becomes increasingly difficult for parliaments to assume responsibility in any meaningful way for the regulation of new technological developments supported by governmental policies. Scientific and technological developments are often of very complex and technical in nature and take place as part of globalized processes where changes occur on a scale that reaches far beyond day-to-day politics. Recent debates and controversies on stem cells, human cloning, genetic testing or nanotechnologies are only a few examples of the difficulties that parliaments face when addressing science and technology developments and related issues.

In this chapter, we discuss how technology assessment (TA) and closely related ('TA-like') approaches can support parliaments in science and technology governance. Alongside Grunwald (2011), we shall argue that TA can contribute to policy making on science and technology 'by integrating any available knowledge on possible side effects, by supporting the evaluation of technologies according to societal values and ethical principles, by elaborating strategies to deal with inevitable uncertainties, and by contributing to constructive solutions of societal conflicts around science and technology'. We shall state that TA is a particularly effective approach to addressing the range of global issues which spill over the borders of nation states, and the chapter calls for parliaments and other policy actors to foster the deployment of TA activities across Europe.

We base our discussion on exchanges made in two parliamentary TA debates that involve parliamentarians and policy makers from across Europe, facilitated by the PACITA project.[1] The aim of these

DOI: 10.1057/9781137561725.0013

debates was to build a common understanding of the role of TA for parliaments in Europe and to discuss further developments of TA activities. Parliamentarians and policy makers who attended the debates stressed the importance of having structured knowledge regarding new technologies that takes into account the scientific aspects as well as the interests and values present in society so as to support processes of policy making. They also defended the pooling of TA efforts across Europe – for instance, through an association that involves a large set of institutions or research groups performing TA (or TA-like) activities. Such an association could carry out concrete activities such as conferences, cross-European projects or exchange programmes for TA staffers, which would constitute an essential step towards the deployment and strengthening of TA policy advice in Europe.

Parliaments and policy advice

The increasing role of science, technology and innovation in Europe has major implications for parliaments with regard to technological developments and/or science-related policies. Parliaments have to regulate the development and use of technological innovations in order to mitigate risks or prevent abuses, but also they also have to set the framework for technological innovation to achieve specific policy goals – for example, health, environment or energy – or to meet public concerns such as security, economic and financial stability or employment. This requires parliamentarians, as well as other policy makers, to achieve a comprehensive view on the issues at stake, taking into account the ethical, legal and societal dimensions of science and innovation. For this, they need to rely on scientific advice that fits their needs and is not influenced by lobbyists and interest groups. In the 1970s and 1980s, members of parliaments made the first calls for TA in Western and Northern Europe. At that time, science and technology were subject to vigorous public debates (e.g. nuclear energy, nuclear proliferation, pollution and so on), and parliaments needed independent and comprehensive analyses and advice on policy options that were based on credible and scientific methodologies. Some 40 years later, these claims continue to be valid, even though the world we live in has changed. Public debate and controversies on science and technology are still present but seem to have waned in intensity (see also Chapter 2). However, the issues in debate are

DOI: 10.1057/9781137561725.0013

more global and complex, and information is moving very fast; together, these make the provision of well-informed and yet independent and structured policy advice critical. René Longet, a former member of the Swiss Parliament, who in the early 1980s initiated the process whereby TA was installed in Switzerland, stated: 'It is a democratic requirement to organize discussions on the ways to manage and guide technological developments for the good of society'.

The importance of scientific knowledge in policy making is of course not new, and it has contributed to the creation of modern states based on rationalization and bureaucracy (Ezrahi, 1990, Latour, 1993). However, the role of science in policy making has long been conceived in terms of a dichotomy between facts and values, wherein science was considered as the domain of facts and causal relationships and politics was the one of values and decisions. This rationalistic model of policy advice, however, comes up against the reality of contemporary policy making. Social studies of science and technology demonstrated that a strict dividing line between facts and values doesn't exist and stress the fundamental uncertainties in science and technology (Latour and Woolgar, 1979, Bijker et al., 1987). As a consequence, policy makers not only need to base their decisions on comprehensive and structured expertise but also need to broaden the scope of the expertise to define policies and regulations stemming from a constructive dialogue between politics, science, stakeholders and society. The rationalistic approach of policy advice – according to which scientists provide facts, politicians add values and bureaucrats implement policies – doesn't match current policy making anymore. What seems to be needed is a space where all involved actors (policy makers, stakeholders and civil society) can be brought together so that their perspectives can inform policy making on issues of science and technology. As stated by Felix Gutzwiller, a member of the Swiss Parliament, 'Technology Assessment is not only about getting expert knowledge, but also about revealing the views of stakeholders and of the general public through participatory methods'. The view of what TA can bring to policy making goes in line with the Beck (1992) and Beck, Giddens and Lash (1994) analysis on the so-called reflexive modernization, which stresses the need to open up political institutions to all actors of society. Policy advice as delivered by TA is not only a way to bring knowledge in parliaments but also a means to foster and facilitate dialogue among conflicting interests and values based on the best available evidence. In that sense, the TA institutions and practices

DOI: 10.1057/9781137561725.0013

that have emerged and developed in Europe may be said to showcase reflexive modernization processes at work (Delvenne, 2011).

Technology assessment for innovation governance

In the tradition of TA, there is a preoccupation with assessing the intended and unintended (adverse) consequences of the introduction of new technologies. This relates to one important area of action for the modern state, which is to mitigate the possible risks of innovation by establishing safeguards and to ensure the safety and quality of products. However, modern states also have the role to drive technological innovation so as to create growth and prosperity and to meet societal needs. In Europe, many high-level policies, strategies and programmes, such as the Europe 2020 strategy, the Horizon 2020 framework program or the Lund Declaration, present science, technology and innovation as central elements to achieving the goals of the the Lisbon Treaty. Such trends clearly affect the kind of policy advice that parliamentarians and other policy makers need: the focus is no longer about mitigating possible risks (risk governance) but about designing innovation so as to avoid adverse impacts (innovation governance). For TA, this implies opening up its traditional risk-based approach and framing its assessment in the wider field of innovation policies.

The approach of Responsible Research and Innovation (RRI) which is currently being developed and fostered by the European Union is regarded as a promising path for supporting the needs of policy makers in innovation governance (Grunwald, 2011, von Schomberg, 2012, Gudowski et al., 2014). RRI refers to 'a transparent, interactive process by which societal actors and innovators become mutually responsive to each other with a view to the (ethical) acceptability, sustainability and societal desirability of the innovation process and its marketable products in order to allow a proper embedding of scientific and technological advances in our society' (von Schomberg, 2013). The various methodologies and tools developed by TA organizations – in particular participatory methods – can certainly contribute to the implementation of the RRI approach in concrete policy-making processes that are related to innovation. Several TA institutes already integrated the RRI approach into their work and conduct projects fostering responsible and sustainable innovation paths that involve science, society and stakeholders. This

DOI: 10.1057/9781137561725.0013

is also the case of the PACITA project, as the 'Scenario Workshops on Tele-Assistance and Future Ageing' aimed at providing input for innovation policies by integrating a wide array of stakeholders so as to meet the societal challenges of an ageing society (see Chapter 7). In such projects, TA fosters a sustained dialogue between research, industry, stakeholders, society and parliaments on innovations and related societal challenges.

Technology assessment in a globalized world

Globalization has broadened the range of issues which spill over the borders of nation states and require international norm setting and regulation. This concerns a wide array of contemporary issues, such as poverty, environmental pollution, financial crisis, organized crime, terrorism and privacy protection. Similarly, scientific and technological developments are increasingly transnational in nature and cannot be addressed at the national level only. The governance of nanotechnologies, for instance, is strongly influenced by supranational institutions – such as the OECD, the European Commission or the European Parliament. In other domains, such as climate change, international organizations such as the United Nations have a strong coordination role in terms of goal settings and action. But this globalization of politics does not mean that nation states are disappearing. Many global issues still need local action and decisions, and they are viewed differently from country to country because of the culturally embedded character of both knowledge and policy (Jasanoff, 2005). For example, several European member states are developing their own policies and regulations relative to nanotechnologies, and recently the European Parliament decided to leave it to each country to decide if they want to authorize the culture of genetically modified organisms (GMOs). In the domain of climate change, it is also up to each country to fix its own objectives and set of actions. Other topics such as ageing society, which many countries have to deal with, also need country-specific solutions, related to the national legal system and cultural characteristics.

Technology assessment has long recognized the importance of addressing the global and cross-border dimensions of science, technology and innovation so as to provide adequate and meaningful advice on the contemporary challenges of our societies. In 1987 the Science and Technology Options Assessment Panel (STOA) was created to carry out expert-based,

DOI: 10.1057/9781137561725.0013

independent assessments of the impact of new technologies and to identify long-term, strategic policy options useful to the European Parliament. The European Parliamentary Technology Assessment network (EPTA) was established in 1990 by TA institutes willing to exchange their practices and to bridge the global dimension of science and technology with the specific context of national policy making. Since its establishment, the network regularly invites parliamentarians from European countries to discuss key scientific and technological trends, and it elaborates reports that synthesize the work of its members on specific science and technology issues.[2] Cross-European projects that are implemented within the PACITA project represent a more structured and institutionalized way of providing cross-border and supranational policy advice to both national parliaments and the European institutions (see Chapter 5 and Part II). In such cross-European projects, a common issue is addressed in several countries through the same questions and with the same methodology, allowing for both a global and local examination. Such collaborative and cross-national approach helps policy makers to look at issues beyond national borders and integrate global challenges into national policy agendas. Findings within the PACITA project also suggest that cross-European projects constitute an opportunity for institutes which are not, *stricto sensu*, TA institutes to join the TA community and develop new skills and new advisory services which are currently not considered in their country.

Putting TA to the political reality test

The PACITA Parliamentary TA Debates were designed to build a common understanding of the role of TA in policy making on science, technology and innovation. The aim was to integrate the views and needs of parliaments in the discussion on knowledge-based policy making in Europe and to reflect on the best approaches to achieve it.

Parliamentarians and policy makers who participated in the PACITA Parliamentary TA Debates have recognized the value of TA to their political work, considering it a democratic tool that besides providing structured knowledge also brings new issues and perspectives into the political agenda and debates. For instance, Maria de Belém Roseira, member of the Portuguese Parliament, told the assembly that 'we [members of parliaments] have to fight blindness when we legislate, we have to have strategic thinking and we need to be aware through

DOI: 10.1057/9781137561725.0013

information. So technology assessment is a very important tool'. Her Austrian colleague Ruperta Lichtenecker shared a similar view and called for 'an open and transparent approach to decision-making in order to improve the quality of decisions reached, to stimulate public debate and to build general awareness on topics that are essential for our future'.

However, the TA approach may compete with other forces that are characteristic of current political decision-making processes. TA operates in a landscape of existent opinions, interests and priorities, and the inputs that it provides for policy making may be drowned out by political bargaining processes and the interplay of various interests, values and strategies. Furthermore, policy makers may select information from TA that supports their opinions and positions rather than using the results of TA to evaluate the available options.

From the perspective of the parliamentarians, another issue to consider when using TA in their work lies in the different time perspectives of cycles in politics and science. Science in general (and TA in particular) is rather well equipped to provide policy advice to decision makers on long-term issues such as innovation strategies or regulation. But matters often arrive without warning on the political agenda for which parliamentarians are expected to react immediately. However, participants of the Parliamentary TA Debates were convinced that the long-term perspective of TA is an essential and unique feature that should be maintained. Several speakers recalled that democracy needs long-term political thinking and that TA is an essential tool to integrate long-term and strategic thinking into politics. According to Joëlle Kapompolé, a former member of the Wallonia Parliament in Belgium, who has been involved in creating a TA office in her region, 'Technology Assessment is the best way to make better decisions for the next generations'.

Reinforcing communication between parliaments and TA

The scientific and political differential processes highlighted by the long-term and comprehensive approach of TA, on one hand, and the constraints of political systems based on representative democracy, on the other, makes it necessary to build permanent and consistent communication between TA organizations and parliaments. It is essential for TA organizations to be aware of the needs of parliamentarians and other

DOI: 10.1057/9781137561725.0013

policy makers, as it is important that policy makers know what technology assessment has to offer them. In that sense, the discussions that took place in Copenhagen and Lisbon during the Parliamentary TA Debates were a unique opportunity for the TA community to hear from the parliamentarians themselves about what their needs are with respect to policy advice on science and technology, as well as for the parliamentarians to get a full picture of what TA offers to policy-making processes and to them personally in their daily work and responsibilities. As such, the Parliamentary TA Debates can be considered as the first step towards an enhanced dialogue between the TA community and parliaments on the contribution of technology assessment to knowledge-based policy making in Europe.

Work still needs to be done to ensure that the nature, methods and effectiveness of TA are better and more widely communicated to policy makers, thus sensitizing them to the benefits of TA and enabling the adoption of TA practices more widely (see also Chapter 9 and Chapter 10). In countries where TA is less developed, the growth of TA practices is often slow, not because policy makers do not really want them, but because TA is not formally part of the decision-making process and may be hence seen as an unnecessary barrier to prompt policy making. Even in countries where parliamentary TA has been institutionalized, its relevance – or even existence – is not necessarily noticed by parliamentarians, which can lead to the closure of productive and successful TA organizations. This is what happened to the US Office of Technology Assessment (OTA), which was shut down in 1995 due to budgetary constraints and bargaining without parliamentarians' noticing it. The same happened to the Danish Board of Technology (DBT) after the 2011 election, but in this case the DBT managed to be transformed into a non-profit foundation. According to Ulla Burchardt, who has chaired the German Parliament's Committee on Education, Research and Technology Assessment and now teaches at the Technical University of Dortmund, 'TA is something apart, for which members of parliaments do not receive any recognition for the next election'. Thus, even though a country may have a long tradition of TA, continuous communication with decision makers is necessary to anchor it in the policy-making landscape and to constantly show its added value to parliamentarians.

But building a common understanding of the role and value of TA for policy making requires more than explaining to parliamentarians what TA is and can offer them. Parliamentarians and other policy makers need

DOI: 10.1057/9781137561725.0013

to be sufficiently involved in TA activities so that they can take owner-ship of the results. For instance, parliamentarians may be involved in setting the agenda for TA activities, may be consulted in the course of the project or may pilot TA activities. In some countries, this link between TA and parliaments has been institutionalized, and if we refer to the TA models presented in chapter one, these institutions are based on strong parliamentarian involvement (see also Ganzevles et al., 2014). This is, for instance, the case of the French OPECST, where the parliamentarians themselves perform TA and their staffers have an auxiliary function; of the German TAB, whose steering committee is solely composed of parliamentarians; and of the English POST (Parliamentary Office of Science and Technology), which is placed directly inside the parliament and works in close contact with MPs. But for many organizations that try to introduce TA in their country, there are no such formal links with parliament. Thus, such links need to be constructed and fostered so that the TA expertise is connected with the political realities and parlia-mentarians get the feeling of owning the TA products. For instance, the participation of parliamentarians from all over Europe in the PACITA Policy Hearing on Public Health Genomics was a unique opportunity for the involved parliamentarians to get a better understanding of what TA can bring them when they have to deal with controversial health tech-nologies (see Chapter 6). This project and other similar projects provide evidence that the ability to build consistent communication channels between policy makers and other relevant actors (e.g. technical experts) is crucial for the effectiveness of TA in policy-making processes. And, on a more general perspective, it offers insights on the type of questions and issues that policy makers are likely to raise and have to face when considering complex scientific and technological developments, which is of great value for the deployment of further TA activities in countries or at the European level.

Parliamentary TA in a context of limited resources

In the current context of financial constraints, most countries are facing economic difficulties and budget cuts, making the public resources required to establish TA practices limited. Therefore, parliaments have to find a reasonable balance between the need for independent policy advice and what a TA unit or 'TA-like' institution could contribute to the

DOI: 10.1057/9781137561725.0013

policy-making process. For instance, parliaments which are currently considering the establishment of a TA unit, but which face budgetary constraints, could consider creating a very small structure (based inside or outside parliament), supported by universities, science academies, research agencies or science foundations. These could support projects that focus on issues of interest for the national political decision-making process, as well as issues of global convergence. The main objective of these projects would be to support members of parliament on policy making and to foster their involvement in TA activities. This work could be supported by fellowships, as in the case of the Parliamentary Office of Science and Technology (POST) in the UK, in which research fellows support the work of the permanent staff.

Another option for countries in which TA is not (yet) well established and is facing budgetary constraints would be to have access to the work done by established TA institutions in other countries. Since many technological issues of interest to policy makers are debated in several countries, some TA groups or 'TA-like' units may 'import' relevant findings made by other TA organizations and analyse them by considering their national context and reflect on the best approaches to start a national debate on the topic in question and involve the relevant stakeholders. According to the resources and TA specific skills available, this option may be achieved by translating TA reports that present, for instance, the state of the art of a scientific field or a meta-analysis of the chances and risks of a given technology, by producing policy briefs on the basis of existing work done by TA institutes abroad and the analysis of the national context and strategic needs of the country, or by initiating a larger process in which local policy makers and relevant national stakeholders would be involved.

Beyond the question of the most appropriate TA institutional model for a specific country, it is important for policy makers to take into account that, while technological innovation is considered a key factor that allows the long-term economic development of a country, TA is uniquely placed to identify strategic options for innovation policies. Moreover, at a time when science and technology are at the centre of growth policies, decision makers need more than ever to rely on tools and approaches that contribute to knowledge-based decision making. This led David Cope, former Director of POST, to state somewhat flippantly: 'If TA is what it claims to be, it is at a time of financial constraints that you need TA more than ever, because TA provides pointers towards how to move out of the period of financial constraints.' Following Cope's

DOI: 10.1057/9781137561725.0013

statement, although the financial context will impose clear limitations to the establishment of new policy-advice entities, TA should be considered a crucial and strategic asset precisely because it analyses the relevant knowledge and information and then integrates it not only in terms of financial investments and economic growth but also from the perspective of desirable or undesirable societal outcomes.

Final remarks: TA bridging national and European debates

As technological developments have the potential to have large impacts on societies, it is very important that they are democratically debated both by parliaments and, more broadly, within society to ensure that their implications are fully understood and evaluated. This is the task of TA, and during the Parliamentary TA debates participants have repeatedly stated the importance of TA to improve the relationship between parliaments and science, but also the difficulties in maintaining TA activities and disseminating this approach throughout Europe. As stated by António Correia de Campos, former member of the European Parliament and chairman of the STOA Panel, 'a good understanding of the interactions between science and society is increasingly important for policy-making in order to mitigate risks, to avoid gaps in regulation, and to increase social welfare, making the most out of future opportunities'.

With the exception of STOA, TA activities are rooted within national contexts: TA or TA-like institutions are supported by local or national agencies, and their outputs are expected to contribute to policy making mainly at the national level. However, scientific and technological developments are driven by global forces, and they have implications beyond national borders. In that respect, TA should be able to create and operate in an environment that takes into consideration both the national (cultural, social and historical) and the European contexts, striking a balance between the skills and strategic needs of individual countries and of the European Union. This is a challenge for TA, but it can also be viewed as a chance. In the case of countries which are currently considering the establishment of a TA unit but face budgetary constraints, the fact that parliamentarians have to deal with similar issues as their colleagues in other countries offers opportunities for resource-effective ways of collaboration. It is also a way to incorporate the global dimension

DOI: 10.1057/9781137561725.0013

of science and technology in the policy advice of TA. The three cross-European projects organized within the PACITA project, for instance, were designed so as a same issue would be addressed in the same way by several national partners. This clearly reduced the costs for the involved partners, but it also contributed to further opening up to supranational concerns and differences among national policies.

In addition to very concrete advisory activities such as the cross-European projects, many other activities could benefit from cross-border fertilization. The Parliamentary TA Debates, for instance, were a unique opportunity for parliamentarians to meet their colleagues from other countries and compare and learn of certain issues discussed in other parts of Europe. Parliamentarians were fully aware of the relevance of bringing TA up to the European scale: in that respect, the creation of a European-wide networking structure (a kind of 'European TA association') would create the ground for the deployment and strengthening of TA across Europe, as several partners would have the opportunity to work together on a same issue and eventually influence European policy making while having specific activities targeted at the national politicians, experts, stakeholders or citizens. Such a network would also act as a capacity building platform, through conferences, thematic or methodological workshops or exchanges of TA staffers. Not only would this enhanced collaboration be effective in contributing to national and European policy making, but as PACITA proved, it would also foster TA skills across Europe that would support broad and long-term strategies for the development of science, technology and innovation.

Notes

1   A first debate was held at the Danish Parliament in June 2012 (Bütschi, 2012), and a second debate took place at the Portuguese Parliament in April 2014 (Bütschi, 2014).
2   See, for instance, the EPTA Briefing note on Synthetic Biology (http://www.eptanetwork.org/documents/2011/EPTA_briefingnote_nov2011.pdf).

DOI: 10.1057/9781137561725.0013

5

Doing Cross-European Technology Assessment

Marianne Barland, Danielle Bütschi,
Edgaras Leichteris and Walter Peissl

▶ **Abstract:** *The authors give a case-based state-of-play account of cross-European TA cooperation in service of national parliaments as well as the European Parliament. Most TA units have formed their role around the specific needs of their national or regional parliaments and other national or regional target groups, making it challenging to shift focus and create new roles for themselves in a European sphere. This article presents recommendations on how cross-European TA can be done in the future with a focus on three aspects of cross-European TA: (1) the added value of cross-European work and lessons from past experiences; (2) the identification of efficient and credible modes of cooperation to conceptualize cross-European TA; (3) the identification of relevant target groups and addressees and the bringing about of impact on the European level.*

Klüver, Lars, Rasmus Øjvind Nielsen, and Marie Louise Jørgensen, eds. *Policy-Oriented Technology Assessment Across Europe: Expanding Capacities.* Basingstoke: Palgrave Macmillan, 2016. DOI: 10.1057/9781137561725.0014.

As a consequence of globalization and European integration, politics is moving upwards, and policy making on many science- and technology-related issues needs a cross-border approach. However, when we look back at the history of European TA, the development and use of technology assessment has been characterized by national and regional efforts, with little capacity for doing cross-European work. As the EU grows, and all European countries become more connected, cross-European TA can contribute to knowledge exchange and capacity building between countries and regions – and as a result provide robust and independent policy advice for European policy makers as well as other traditional target groups in the national context. Issues related to science and technology are often discussed at a European level, and it seems only natural that these discussions should inform each other and contribute to a broader knowledge base for decision making – whether on a regional, national or European level. The PACITA project, therefore, aims at encouraging practices of cross-European TA in order to strengthen the knowledge base for policy making in Europe.

In this chapter, we discuss the challenges of doing cross-European TA in practice and the framework conditions for using TA transnationally at the European level. In the introduction to this book, we have seen how cross-European TA may fit within existing frameworks for European cooperation. This chapter supplements the introduction by providing an 'on-the-ground' account of the practical and organizational work that it takes to carry out TA projects in trans-European cooperation. We base our discussion on case studies of previous cross-European projects and on new experiments carried out within the PACITA project, all of which have produced important insights on the added value of cross-European TA and how it may be done in the future. These insights show the diversity and inclusiveness which have become characteristic for cross-European projects. Cooperation and communication across borders not only provide knowledge exchange but create arenas and networks for knowledge production and policy learning among European member states and European institutions. Participation in cross-European projects will therefore benefit society's ability to comprehend issues related to science and technology and at the same time open up the process of policy making, making it more understandable and accessible for European citizens. Our findings, however, also show that cross-European TA has so far been conducted on a project-by-project basis, which means that new cooperation forms and capacities have to be established for each

DOI: 10.1057/9781137561725.0014

project. There is therefore a need to develop a European platform that would ensure support for cross-European projects, with regard to both financial and human resources.

Cross-European technology assessment: current situation

Several research projects and reports have documented the activities and methods of TA in Europe,[1] but few of these have discussed cross-European cooperation and how this can be done in the best possible way. The PACITA project had a goal of making recommendations for the future of cross-European TA, based on lessons learned from past examples of cross-European projects as well as research done in the PACITA project.

Although a STOA report (Enzing et al., 2012) from 2012 describes cross-European TA as limited, there have been several European and international TA projects over the years. Experiences and lessons learned from these projects give important input for further development of work modes, methods and funding schemes. The PACITA project has conducted a number of case studies with the aim of identifying the added value of the cross-European approach, as well as identifying some of the barriers and challenges related to these types of projects.

The EPTA (European Parliamentary Technology Assessment) network is an example of an existing network of European PTA units. Together, the partners of EPTA aim at making TA an integral part of policy consulting in parliamentary decision-making processes around Europe. EPTA has initiated and organized several cross-European projects. These projects[2] are always funded on the partners own budget, as the network itself does not have any resources. This funding scheme creates certain limitations in the project design, and the method in EPTA projects has over the last years been limited to distributed desktop research, in which all partners write a state-of-the-art chapter from their country/region on a given topic and present policy options. The contributions are then collected and presented in a common report, opened by a short introduction written by the project coordinator. There is rarely any in-depth cross-European analysis of the national contributions, but taking their minimal resources into account, these projects have a good record of accomplishment. Feedback on the joint EPTA projects shows that parliamentarians appreciate seeing how other countries deal with the same challenges as themselves.

DOI: 10.1057/9781137561725.0014

Another type of projects is funded through the European Union's Framework Programs,[3] like the PACITA project. The projects are based on project calls from the European Commission and cover a broad spectrum of topics. These projects have dedicated budgets that make it easier to use more demanding methods than the EPTA projects. This can include methods that involve citizens or stakeholders in addition to more traditional desktop research. A consortium in these projects often involves several types of partner institutions (universities, NGOs, research institutes, TA institutions etc.).

A third type of project[4] is commissioned by STOA (the TA unit of the European Parliament) and carried out by members of European Technology Assessment Group (ETAG) or other consortia. These projects have both a dedicated budget and pre-defined target group in STOA. The projects cover a variety of topics and use mostly desktop research and expert hearings as methods. One challenge with commissioned projects is that it can be difficult to identify the most relevant scope for policy makers when taking on topics where extensive research has already been done. That the project is scientifically 'less free' when the project is commissioned by a 'client' can also be challenging.

The PACITA experience

From the pool of previously conducted TA projects, there are several types of projects and consortia which differ with regard to funding schemes, methods, target groups and project designs. PACITA organized three example projects, aiming to produce relevant policy advice at national, regional and European levels. The projects also aimed at enhancing the capacity of technology assessment in Europe by including both experienced institutions and 'newcomers' in the field of TA. On a more practical side, the projects functioned as an introduction and as training for TA practitioners involved in the PACITA project.

The three example projects took on three of the Lund declaration's 'grand challenges', using different methods and involving different types of actors:

While scenario workshops and citizen summits are quite established methods at the European level, it was the first time that the Future Panel was used in a cross-European manner. This 'methodological experiment', together with the two more established methods, has given important

DOI: 10.1057/9781137561725.0014

TABLE 5.1   *Overview of PACITA example projects*

Topic	Method	Involved actors
Personal health genomics	Future Panel	Parliamentarians and experts
The future of ageing	Scenario workshops	Stakeholders
Sustainable consumption	Citizen summits	Citizens

insights on how to organize successful cross-European TA projects (see Part II of this book).

One of the challenges related to the Future Panel method, was the need for long-term commitment by parliamentarians. Earlier experiences with the Future Panel method on the national level have involved parliamentarians who have been appointed to the Future Panel by their parliament (Krom and Stemerding, 2014). A more direct link to the national parliaments (and not only involvement of individual parliamentarians) makes a clearer mandate for participation in the project, and it will probably make it easier for parliamentarians to commit to the project. The two other example projects had a single national event as the main activity. The activity demanded some preparation by the participants (reading information material or scenarios), but it demanded no long-term commitment to the project.

One might argue that by doing such national events, the cross-European element is put in the background. But seeing that both the citizen summit and the scenario workshop had a common European starting point for the discussions,[5] the participants still got the feeling of being part of a European project. Knowing that there are others having the same discussions, following the same method, somewhere else in Europe was acknowledged and appreciated by the participants. In miniature, the deliberative fora that were created within the projects seemed to engender an experience of European citizenship solidly rooted in national communities. The results from these national events were gathered in European synthesis reports, bringing the results from the national to the European level.

In addition to the policy recommendations produced by all three example projects, an important result is the added value for the TA community. Focus on method training gives all of the involved partners a strong foundation to further use these methods also after the end of the PACITA project, and it enhances the capacity of the involved institutions.

DOI: 10.1057/9781137561725.0014

Barriers to cross-European TA

Although there have been a number of cross-European projects that have been conducted over the years (as described above), one cannot speak of regular cross-European TA having been done.

National vs European commitments

However, a tension might occur for each individual organization between doing national projects and participating in European projects. This tension may act as an obstacle for developing cross-border collaboration. Easing this tension might be a factor that can lower the threshold for TA institutions to engage in cross-European TA. Most of the existing TA institutions have their mandate mainly focused on the national and regional spheres. Some have an identified task to 'watch trends in science and technology' (on both the national and the international level) (Ganzevles and Van Est, 2012), but none have international cooperation as a defined task. Identifying and understanding the added value in cross-European projects may help to open up and stimulate more cooperation and at the same time justify international cooperation with regard to mandates and resources, without stealing attention away from national working plans.

Finding a European audience

One of the main characteristics of the traditional TA units has been their strong connection to parliaments (see also Chapter 1). This relationship has often been institutionalized either by organizing the unit inside parliament or by stating this relationship in the mandate of the institution. Some 40 years later, the audience of TA or TA-like institutions is wider and includes all actors involved in policy making – that is, members of parliament, but also governmental representatives, civil society and even the scientific community. However, these actors are mainly nationally based, showing that the audience of TA lies within usually national (or regional) frontiers.

When TA activities take place at the European level, it becomes more difficult to create permanent relationships with addressees and potential target groups than in national projects. In national contexts, there exists a defined public sphere, although there is no clearly defined 'European public'. One possible approach is to have a broader view of addressees and

DOI: 10.1057/9781137561725.0014

target groups when working at the European level than at the national/ regional level. If the goal of TA is to give input for evidence-based decision making, it might help to widen the definition of who decision makers in fact are. In the European context, the European Commission and the European Parliament play important roles as policy makers. But Europe is multifaceted and consists not only of the European Union; many others (lobbyists, NGOs and the media) take part in decisions and hold power in important discussions about the policy issues and options. Therefore, all those organizations and institutions can be potential target audiences for cross-European TA, on the European as well as the national level. Nations are an important part of, and often the operative level, European policy making. They should, therefore, also be an addressee of cross-European project results. In order to reach such an audience, focus should be on communication efforts and on forming clear and targeted policy advice.

One important audience is the TA community itself. Results from successful cross-European projects can be used at the national level from institutions not involved in the specific project and also as an encouragement for participation in future cross-European work. This would contribute to a bigger pool of evidence of cross-European work – hence raising the legitimacy and the trust in a cross-European approach and in TA methods.

Benefits of cross-European TA

Based on the challenges related to European projects, it is important to identify the defining elements of cross-European TA and to understand what makes technology assessment an important contributor for policy advice in Europe.

For society

The emerging technologies debated in different countries are more or less the same. But contexts and timing of discussions, and the shaping of technologies, will differ nationally. Thus, cross-European TA can contribute to agenda setting and provide policy support at the European level and at the same time inform national science and technology discourse. This has already been identified in the area of European science policy, moving from 'science in Europe' to 'European science' (Nedeva and

DOI: 10.1057/9781137561725.0014

Stampfer, 2012). Focus has moved from coordination of national projects to the development of a more integrated, pan-European science base. When topics are relevant across borders, it's reasonable to think that it would be more effective to make projects on a cross-European basis rather than have every TA unit do similar projects in their country/region.

For parliaments

In the 1970s, when TA started to get institutionalized in Europe, the influence of the American tradition of TA was evident. However, as argued by Norman J. Vig (2000), the European approach to TA turned out as more of a democratic project than it had been in the US, where the focus had mostly been on creating an informed policy debate on science and technology issues. Introducing TA in the diverse and culturally varied Europe, TA became a strong instrument in the democratic process, providing independent and thorough advice for parliaments, based on participation of a broad group of actors. This is also one of the reasons for the survival of these organizations, Vig argues: they have proved useful for parliaments.

For TA institutions

PACITA is in itself a good example of how TA institutions benefit from doing cross-European projects. PACITA strengthened the ties between the existing TA units, and it also helped establish a strong base for further institutionalization of new initiatives in Europe. Doing PACITA's three example projects proved that participation in cross-European projects is highly productive from a practitioner's point of view. The cooperation provided institutional learning and an exchange of experience between TA practitioners, and the hands-on experience from the projects created enthusiasm for TA both among the participating institutions who were new to the field and among the policy makers who received the results.

Requirements for realizing cross-European TA

An essential element of TA is the notion of independence. This refers to the independence of TA institutions from stakeholders' interests and influence, as well as the independence from funders and policy makers

DOI: 10.1057/9781137561725.0014

themselves. Independence is important to maintain the TA institution's credibility, and it will strengthen the reputation of TA in Europe at a more general level. Giving well-founded and independent advice is one of the main strengths of TA, compared to policy advice from NGOs and lobby groups, who have their own interests in mind.

Future cross-European TA initiatives should be both inclusive and diverse. Acknowledging that others see similar challenges but deal with them differently can lead to knowledge and new perspectives. Cross-European TA can contribute to agenda setting and policy support at the European level and at the same time inform national science and technology discourses. The PACITA project had a variety of partners, not only traditional PTA institutions. The diversity of the consortium combined with the cultural backgrounds of the countries and regions involved created a learning process for all partners – and contributed in new knowledge production for policy makers. However, there will always be challenges related to cross-European participation and national financing. Seeing that the financial situation of the different national and regional institutions varies, it is difficult to ensure the diversity of TA on the European level.

In the last few years, the field of TA has changed. Several institutions have been transformed and reorganized, and one can see a need to broaden the scope of European TA, from purely parliamentary TA (PTA) to forms of TA that approach policy making in a broader way. PACITA's efforts in expanding TA throughout Europe highlight the democratic approach to TA that is taken in Europe, and the introduction of TA in new countries, regions and cultures will add value to policy makers and the TA community. A more permanent and stable presence of TA at the European level also will serve as important support for TA initiatives in the future.

Creating a permanent and stable presence of TA on the European level, and making it easy and desirable for TA institutions to participate in cross-European projects, demands more systematic funding than is provided today. The experiences from previous TA projects might seem to argue that as long as there are funding mechanisms available, such as the EU framework programmes, then cross-European TA will continue to exist. However, there is a strong belief that cross-European TA can grow even stronger if there is more systematic financing for cross-European cooperation, which is not limited to individual projects. A continuous presence, such as in the format of a TA Platform, will make

DOI: 10.1057/9781137561725.0014

a stronger impact than individual national institutions coming together for projects now and then (see also Part III of this book).

There has been an increase in cross-European initiatives in the field of TA. This is reflected in the number of projects, the number of participants and the involvement of new countries and institutions. The TA community in Europe has historically been oriented towards producing policy advice for national and regional parliaments. Because of the shifting landscapes in Europe, it makes sense to extend the addressees to a wider group of policy makers. This move will give greater opportunities for making an impact in a wide range of policy processes. At the same time, it will open the field of TA to participation of a broader group of institutions, not only the 'traditional' institutions doing parliamentary technology assessment. A variety of institutions are now active in the field of TA in Europe. They all have to find their own strategies for how to be agile and flexible enough to participate at European level, yet at the same time deliver results to the national policy makers.

The three example projects organized during the PACTIA project have provided insights on three of the grand challenges that our societies will face in the coming decades. The approaches made available through technology assessment has produced important input for policy makers and also demonstrated the important role that institutions for technology assessment can play at the national and the European level. Experiences from these three projects highlight especially two methods that work well on the cross-European level: citizen summits and scenario workshops. Having a common starting point (information material or future-oriented scenarios) in national activities gives the approach a common thematically starting point, but it also allows room for the cultural and social differences in countries and regions. This also produced output that is valuable for national, regional and European policy makers.

Final words: making an impact

In the end, the goal of TA is to make an impact on policy making. And its 'impact' can be manifold. It can contribute to bringing new or independent knowledge to science and technology themes or to the related societal aspects in policy-making processes; it can contribute to agenda setting; it can act as a mediator or facilitator between stakeholders; or it can lead to new policies or regulations being made (Decker and Ladikas, 2004).

DOI: 10.1057/9781137561725.0014

Even though some institutions have formal relationships with important policy makers, these policy makers are not demanded to act upon the advice coming from the TA community. One of the main characteristics of TA is its way of bringing together knowledge from a broad group of actors into the production of independent and well-grounded policy advice. By using existing as well as by further developing traditional methods, the TA community should strive to enhance evidence-based policy making at the national, regional and European levels.

The developments and discussions related to science, technology and society move forward with increasing pace. In order to advise policy makers on these developments as they unfold, TA institutions must be present and in contact with their target groups at all levels. Seeing that these developments happen on a European level and an international level, the need for cross-European TA is evident. Cross-border knowledge exchange and learning is highly relevant for policy makers in our societies today, and cross-European TA represents one way of making this happen.

Case studies based on the following projects:

▸ Energy transition in Europe (2007)
▸ Genetically modified plants and foods (2009)
▸ ICT and privacy in Europe (2006)
▸ Energy transition in Europe (2007)
▸ Genetically modified plants and foods (2009)
▸ ICT and privacy in Europe (2006)
▸ Challenges of Biomedicine (2007)
▸ CIVISTI (2011)
▸ Meeting of Minds (2006)
▸ Study on Human Enhancement (2009)
▸ Nanosafety (2011)
▸ Technology Options in Urban transport (2011)
▸ PACITA example projects: Personal Health Genomics, the future of ageing and sustainable consumption (2013–15)

Notes

1   For example, EUROPTA (2001) and the TAMI project (2004).
2   Examples from the case studies include 'Energy transition in Europe' (2007), 'Genetically modified plants and foods' (2009) and 'ICT and privacy in Europe' (2006).

DOI: 10.1057/9781137561725.0014

3   Examples from the case studies include 'Challenges of Biomedicine' (2007), 'CIVISTI' (2011) and 'Meeting of Minds' (2006).
4   Examples from the case studies include 'Study on Human Enhancement' (2009), 'Nanosafety' (2011) and 'Technology Options in Urban transport' (2011).
5   Information material and short films for the citizen summit, as well as scenarios in the scenario workshops.

DOI: 10.1057/9781137561725.0014

Part II
Exemplifying Cross-European Technology Assessment

DOI: 10.1057/9781137561725.0015

6

The Future Panel on Public Health Genomics – Lessons Learned and Future Perspectives

André Krom, Mara Almeida, Leo Hennen, Edgaras Leichteris, Arnold Sauter and Dirk Stemerding

▶

Abstract: *Krom et al. give an in-depth account of a methodological experiment carried out in the PACITA project, namely the application in a cross-European context of the Future Panel method. Focusing on the complex issue of genomics and its potential use in public health care, parliamentarians from different countries were gathered to learn about and debate this far-reaching field of research in order to create a foundation for proactive policy formulation. The authors analyse and evaluate the project setup and argue that while further development and institutional is necessary to make similar future projects reach their full potential, the project nevertheless exemplifies the practicability and value of applying previously nationally contained TA methods in a cross-European setting.*

Klüver, Lars, Rasmus Øjvind Nielsen, and Marie Louise Jørgensen, eds. *Policy-Oriented Technology Assessment Across Europe: Expanding Capacities*. Basingstoke: Palgrave Macmillan, 2016. DOI: 10.1057/9781137561725.0016.

 DOI: 10.1057/9781137561725.0016

Technology constantly pushes the bounds of what medical care can achieve and at what cost. Although medical care is a highly expert-driven field, parliamentarians and government decision makers nevertheless become involved in shaping medical innovation through funding decisions and framework regulations. If such interventions are to be both legitimate and effective, they must be made on the basis both of sound evidence and of open dialogue regarding possible pathways. Designing processes to ensure such quality in policy making is a key example of the role that technology assessment (TA) institutions can play as mediators between science and policy. To exemplify this role to European policy makers, PACITA carried out an experiment in cross-national policy dialogue on Public Health Genomics (PHG).

PHG is often understood as the responsible and effective translation of genome-based information and technologies (GBITs) into health-care practices. It is regarded as a central future perspective for the medical system. According to some experts, PHG will make health care truly personalized, predictive, preventive and participatory. However, there is still a high degree of scientific uncertainty about what PHG can actually deliver. There are also far-reaching ethical, legal and socioeconomic questions related to GBITs. Therefore, an in-depth societal and political debate on PHG is of fundamental importance for the future health-care system.

TA has already played an important role in the public and political discourse in many countries, by systematically collecting inter- and trans-disciplinary knowledge and by stimulating and organizing debate between different stakeholders. Given the rapid scientific progress and many challenges for policy making in the foreseeable future connected to PHG, an expert-based methodology – the Future Panel – was chosen. The central idea behind the Future Panel method is to connect the scientific and the political discourse in a new and constructive way. In general, the method is well suited to far-reaching topics that require central political initiatives and action and where there is a desire to act proactively. The method had originally been developed and applied in a national context. In this project, the Future Panel (FP) was formed by parliamentarians from different European member states and the European Parliament with a specific responsibility for health policy. This was a methodological experiment because the FP method had to be adapted to a cross-national context.

As an *example* project, the FP on PHG succeeded in contributing to the central aim of PACITA – to induce mutual learning on setting

DOI: 10.1057/9781137561725.0016

up support platforms for knowledge-based decision making among the European countries involved. The project also managed to provide relevant input for policies on Public Health Genomics in terms of an overview of state of affairs and policy options. Developments in PHG hold the promise to be beneficial for individuals and to promote public health. However, given a range of uncertainties and ambiguities related to GBITs, the responsible introduction of GBITs in health-care systems requires an incremental approach.

As a methodological *experiment*, the project did not meet all of its objectives, including the aim to connect the scientific and political discourse on Public Health Genomics in a new and constructive way. Due to the complexity of the topic and the specific restriction of time and resources, detailed discussions of options for policy intervention and regulation of existing practices and regulatory stipulations for different fields of application were not possible. Through its broad approach, however, the project and its documented outcomes are useful to raise sensitivities for problems to be expected and thus can serve as a starting point for a more detailed evaluation of single GBIT applications and health-care practices on the European level and the national level.

Background

The aims of the demonstration project were to provide a concrete and policy-relevant example on EU-level coordinated parliamentary TA by:

▶ giving input to policy making on policies on Public Health Genomics, in terms of an overview of state of affairs and policy options;[1]

▶ establishing a national/regional-level and EU-level experience with a coordinated expert-based TA method that involves parliamentarians;

▶ doing this in cooperation with decision makers on the national/ regional level and the EU-level, in order to create experience on, and thereby mobilization around, the use of such methods among the main users;

▶ doing this in cooperation with the scientific community on Public Health Genomics in order to create learning and mobilization on the potential of expert-based policy making facilitated by TA specialists; and

DOI: 10.1057/9781137561725.0016

▸ involving countries that have not established such institutions and methods directly in their work, in order to build capacity, create learning and mobilize the actors.

The idea of installing a panel of parliamentarians to discuss long-term political issues related to developments in science and technology was not new. An example of an earlier and comparable initiative is the Finish Committee for the Future. Based on parliamentary proposals going back to 1986, a Committee for the Future was appointed in 1993 on a temporary basis. In the year 2000, the Committee received permanent status.[2] Building on the Finish experience, the Danish Board of Technology developed the Future Panel method. This method involves a temporary panel, typically for a period of 1½–2 years, the activities of which revolve around intensive collaboration between the Future Panel and invited experts from relevant practices related to the topic at hand.

Like the Danish Future Panel method, the PACITA Future Panel involved a temporary panel of parliamentarians and the collaboration of the Future Panel and invited experts. Important differences were that the project on Public Health Genomics involved a *cross-national* Future Panel, that the interaction between the Future Panel and the invited experts was less extensive and that there was *no institutional link* between the project and the respective parliaments of the FP members: they were invited as individual members of parliament. This meant that the method had to be adapted for use in a cross-national context. In a sense, then, the 'Future Panel on Public Health Genomics' was a methodological experiment.

The Future Panel project: process, participants and outcomes

The Future Panel project on Public Health Genomics consisted of three stages. In the first stage, the precise scope of the project was defined during a kick-off meeting that involved the Future Panel, which resulted in a list of policy issues that were identified as most relevant for further investigation. During the second and main stage of the project, which took a full year, policy issues and options for public health genomics were discussed and elaborated in different expert working groups (EWGs) and in a policy options workshop. The final stage was a Policy

DOI: 10.1057/9781137561725.0016

Hearing in which the Future Panel discussed the main outcomes of the project with invited experts.

The main target group of the project was the Future Panel, consisting of parliamentarians with a specific responsibility for health policy. The panel had four members, who represented different parties in the political spectrum, including one member of the European Parliament and three members of national parliaments (Denmark, Portugal and Switzerland). The main role of the FP was to co-define a research and policy agenda at the start of the project and to discuss, during the final Policy Hearing, the issues and options articulated by a range of experts on different aspects of PHG who were involved in the course of the project.

The project was carried out by a *task team* of TA practitioners from the four countries involved in the PACITA consortium.[3] As in all subprojects of PACITA, partners were from both countries with and countries without established institutes for (parliamentary) technology assessment (see Table 6.1). A group of five external experts on different aspects of public health genomics was involved as a *steering group* to assure the high quality of all project activities. Four international *expert working groups* were responsible for the investigation and articulation of policy issues and options for public health genomics in a year-long process of collaboration with the task team and the expert steering group.

Stage 1: defining an evidence-based policy agenda

As an expert-based methodology, the Future Panel on Public Health Genomics was based on the assumption that policies relating to future developments in this field should be evidence based. 'Evidence' should be taken in a broad sense here: the issues raised by the introduction of genome-based information and technologies in future health care involve not only complex scientific questions but also a history of controversial ethical, social and legal debate concerning highly sensitive areas of medical care, such as prenatal diagnosis and genetic screening. Four international EWGs were composed of experts on precisely these issues. The Future Panel had a pivotal role at the start of the project in identifying the issues that would require further research, deliberation and political action: to ensure the political relevance of the expert-based analysis and policy options to be deliberated in the final policy hearing. During the kick-off meeting of the project, these issues were defined in a discussion with the steering group and task team, resulting in a research

DOI: 10.1057/9781137561725.0016

and policy agenda that raised questions that could serve as input for the ensuing investigations in the four expert working groups.

Stage 2: Exploring the field

With this research and policy agenda as a starting point, the evidence produced by the expert working groups during the second stage of the project covered not only technical state-of-the-art scientific knowledge but also a broad range of other relevant issues raised by developments in the field of public health genomics. The task of the working groups was to produce twenty-page reviews of: (1) the state of human genome research and its prospects for future medical applications in public health genomics; (2) issues of quality assessment relating to the clinical validity and utility of genome-based medical applications and practical experience in public health genomics; (3) the possible economic and structural effects of public health genomics on the public health system; and (4) the ethical, social and legal aspects of public health genomics. In reviewing these different topics, the expert working groups not only engaged themselves with the Future Panel policy agenda in more or less direct ways but also *reframed* this agenda by putting the issues in a broader context of current and potential future developments and challenges in the field of public health genomics. Based on this review, the role of the EWGs further included the articulation of policy options suggesting different ways in which policy makers might deal with the issues raised by future prospects in public health genomics.[4]

The efforts of the expert working groups were coordinated by the task team members, who also had the responsibility to summarize the four working group reports in an expert paper that described in a concise and accessible way the challenges and policy issues that were identified by the experts as most salient and urgent.[5] The expert paper was the central input for the policy options workshop.

The policy options workshop brought together experts from the four working groups and members of the expert steering group and task team, allowing the project participants to further increase the focus of their main findings and to 'translate' into policy options the rather divergent perspectives on public health genomics represented in the project. The results were integrated in a policy brief that served as the main input for the concluding policy hearing.[6]

DOI: 10.1057/9781137561725.0016

TABLE 6.1   *Items highlighted in Policy Brief on Public Health Genomics*

Issues related to medical genomics research
Data sharing and intellectual property
'Big data' security and privacy
Quality assessment

From research to clinical practice
What to screen for and when
Patients' rights and professional responsibilities
Informed consent and service provision

Governance in public health genomics
Need for an incremental and programmatic approach

Stage 3: a new policy agenda?

During the final policy hearing, the Future Panel again played a pivotal role. The hearing was organized as a public meeting in which the Future Panel had the opportunity to discuss with three panels of experts the main items highlighted in the policy brief (see Table 6.1). The aim of the policy hearing was to provide more fine-grained clarifications and suggestions related to the policy questions and issues that were formulated by the FP members at the start of the project. In this way, the FP members would gain a better understanding of the issues involved. Providing information that takes into account the different views on public health genomics would support the FP members in their work in parliament.

The Future Panel as a TA demonstration project – main achievements and implications

As a TA demonstration project, the Future Panel on Public Health Genomics did quite well. To start with, it successfully contributed to the central aim of PACITA, which is to induce mutual learning in support of the establishment of platforms for knowledge-based decision making among the involved European countries (in this case Germany, Lithuania, Portugal and the Netherlands). One example of this has already been mentioned, namely the fact that at the start of the project none of the task team partners had prior experience with the Future Panel method. Over the course of the project, all partners gained experience not only in

DOI: 10.1057/9781137561725.0016

actually applying the method but also in adapting the method and applying it in an entirely new context: a clear example of *mutual* learning. Another example is the fact that the Portuguese partner Instituto de Technologia Quimica e Biologica (ITQB), who got involved in PACITA as a so-called non-PTA country, is now a participant in another TA project that relate to public health genomics, focusing on the 'genetics clinic of the future'.

The project also provided relevant input for policies on public health genomics in terms of an overview of the state of affairs and policy options. It succeeded in involving a broad range of European genomics experts as members of the Working Groups. For instance, interim results of the project have been presented during a satellite meeting of the 2013 conference of the European Society for Human Genetics.[7] Policy makers and practitioners from the countries that were involved were provided with the best available expert knowledge on GBITs and could gain practical experience with TA as a practice of democratic and transparent knowledge-based policy consulting. The complete interactive exercise of Expert Working Groups, Policy Options Workshop and stakeholder consultation support the notion that developments in public health genomics hold the promise to be beneficial for individuals and to promote public health. However, a crucial insight from this process is also that, given a range of uncertainties and ambiguities, the responsible introduction of GBITs in health-care systems requires a careful step-by-step approach that involves a broad societal and political debate about the direction in which health-care systems should develop.

The Future Panel process highlighted two major shifts connected to developments in public health genomics that challenge traditional boundaries in health care. First, the introduction of GBIT in health-care systems challenges the boundary between research and clinical care. It entails complex data flows that raise a number of issues relating to infrastructure demands, intellectual property, data security, tensions between the needs of research and the needs of the individual, patient rights and professional responsibilities, and the potential feedback of (re) analysed data. Second, the introduction of GBIT in health-care systems challenges the boundary between clinical care (particularly diagnostics) and screening. Both diagnostics and screening generate potentially large amounts of information about an individual's genome and raise new and challenging issues concerning quality assessment and how to deal with unsolicited information that might result from these tests. These issues could arise in a variety of health-care settings as whole genome

DOI: 10.1057/9781137561725.0016

sequencing tests find further application in established and new practices of screening. Consequently, the responsible introduction of GBITs in the health-care system requires an early dialogue in which these stakeholders are actively involved.

The ambition of the project was to deal with the full scope of possible future applications of GBITs, such as pre-implantation and prenatal genetic diagnostics, new-born and adult screening programmes, and whole genome sequencing for general medical services. This broad scope was indispensable for an evidence-based evaluation of the pros and cons. The timespan of the project, however, did not allow for detailed discussions of options for policy intervention and regulation or of existing practices and regulatory stipulations for each of the fields of application. Also, a more in-depth analysis of the state of practice in the different countries involved was not possible. Through its broad approach, however, the project has helped to increase stakeholders' sensitivity to foreseeable problems and thus can serve as a starting point for more detailed evaluations of single applications of GBITs and health-care practices on the European level as well as on the national level.

The Future Panel on PHG as a methodological experiment

Up until the PACITA project, the Future Panel method had been used twice by the Danish Board of Technology (DBT).[8] Methodologically, there were clear similarities between the design of the 'original' Future Panel (OFP) as developed by the DBT and the PACITA Future Panel (PFP). Both the OFP and the PFP lasted approximately 1½ to 2 years and started with an introductory seminar in which the Steering Group and Future Panel met for the first time to jointly determine the focus of the project. Like the OFP, the PFP aimed to gather existing knowledge on the central theme in connection with debate and assessment, to create an overview and elucidate the political tasks connected to the theme. Again, like the OFP, the PFP relied heavily on the input of experts to feed into the policy-making process.

However, there were also important differences between the original Future Panel and the PACITA variant. For the purposes of this chapter, we will mention five of them that contributed to the project being a methodological experiment.[9]

DOI: 10.1057/9781137561725.0016

▶ First, while the OFP was developed for and applied in a national context, the PFP involved adjusting this method to and applying it in a cross-national context. It was in this cross-national context that the TA demonstration had to contribute to the broader aims of PACITA: by establishing a national/regional-level and EU-level experience with a coordinated expert-based TA method that involved parliamentarians; by doing this in cooperation with decision makers on the national/regional level and the EU-level, in order to create experience on, and mobilization around, the use of such methods among the main users; by doing this in cooperation with the scientific community on public health genomics in order to create learning and mobilization on the potential of expert-based policy making facilitated by TA specialists; and by involving countries that have not established such institutions and methods directly in their work, in order to build capacity, create learning and mobilize the actors.

▶ A second important difference between the OFP and PFP was that in the OFP panel members were *appointed* by parliament, thereby forging a strong institutional link between parliament and the project. In the PFP, on the other hand, individual members of parliament were invited by the PACITA consortium. In other words, in the OFP, there was no institutional link between the respective parliaments of the Future Panel members and the project.

▶ As a result, and this is the third important difference, the work done by the PFP worked at a greater distance from actual political committee work compared to the OFP. Typically, work done by the OFP can be regarded as provisional political committee work.

▶ Fourth, the OFP and the PFP differed with regard to the political representation in the Future Panel, both with regard to the political spectrum and the parliamentary committees involved. In the OFP, all political parties were represented, as well as a wide range of political committees. This was not the case in the PFP. There was some political diversity, but not all political parties (from all participating countries) were involved. In addition, members of the PFP were all connected to a parliamentary committee with a special responsibility for health-care policy.[10]

▶ Finally, there was an important difference between the OFP and the PFP concerning the number of public hearings that were

DOI: 10.1057/9781137561725.0016

organized as part of the project. Whereas the OFP typically involved four public hearings, the PFP involved one public hearing, complemented by the possibility of consulting the FP members on an ad hoc basis.

Lessons learned and future perspectives

Based on our experiences with the project, we will now present a number of lessons learned about the Future Panel method as a model for evidence-based and anticipatory TA in a broad international context. With these lessons, we would like to address first of all policy makers and civil servants wanting to support cross-European TA.

Lesson 1: Establish a connection with parliaments and/or ministries, in addition to their respective individual members

Contrary to the standard model, Future Panel members in the project on public health genomics were not appointed by parliament(s) but invited by the PACITA consortium. More specifically, the members were (primarily) invited as individual members of parliament based on their particular individual expertise. In addition, the experimental character of the project entailed that the project activities were not directly tied to an explicit mission by a policy-making body. This meant that the work of the Future Panel and the expert working groups started at a greater distance from parliament compared to the standard model. One of the positive outcomes of doing cross-European TA is to provide an opportunity to debate specific issues which are not on the front line of national political discourses but which are in need of urgent consideration and reflection in a European context. As noted, the members of the Future Panel indicated that a possible action following the final policy hearing would be to present the issues discussed in their respective parliaments. Thus, the function of establishing more direct links to national parliaments would be to attain a more clear 'mandate' to offer policy options – not to individual members of different parliaments only, but to their respective parliaments as well.

Parliaments may have less policy-making power in some countries than they do in other countries. Moreover, experience with evidence-based policy making may be concentrated not in parliament, but in the government or the ministries. If the aim of a project is to promote and

DOI: 10.1057/9781137561725.0016

to mobilize experience with evidence-based policy making on a certain topic, then at least with regard to these countries, we would recommend not to focus exclusively on parliamentarians but to invite policy makers from the government and/or ministries as well.

Lesson 2: Establishing a solid evidence base for policy making requires an iterative process that involves direct contact between all actors directly involved in the project

By organizing multiple public hearings, the standard model automatically allows for an iterative process that involves direct communication between the Future Panel and the experts, and between the Future Panel and the steering group. At the start of the PACITA Future Panel, it was indicated that the panel could be consulted during the process on an ad hoc basis. Such consultation was done once, allowing the steering group and the expert working groups to receive feedback on the draft reports of EWGs 1 and 2. However, organizing the contact in this ad hoc way meant that this round of consultation was positioned as something extra, not as an integral part of the process. Moreover, apart from the concluding policy hearing, communication between the FP and the experts in the PACITA project was always mediated by members of the task team. As a result, the project allowed for relatively few opportunities to check whether there was an adequate match between the policy issues and questions raised by the Future Panel, on the one hand, and the findings from the expert working groups, the expert paper and the policy brief, on the other.

Explicitly building an iterative process into the project design would also increase the possibilities to map and to manage mutual expectations. For instance, feedback from the Future Panel after the policy hearing made clear that some members would have expected more practical answers to the questions and issues that the panel formulated at the start of the project. On the other hand, evaluation of the expert working groups showed that not having a clear mandate to offer policy-making solutions raised questions pertaining to the role of the EWG's and may have affected the motivation of individual EWG members to articulate and reflect on particular policy options.

We highly recommend, therefore, to include in the project design of the Future Panel method, an iterative process that involves direct contact between all involved in the project: (1) between the Future Panel and the experts involved; (2) between the Future Panel and the steering

DOI: 10.1057/9781137561725.0016

group; (3) between the members of the Future Panel; and (4) between the experts from the different expert working groups. Especially in the context of cross-European TA, this will require considerably more time and a larger budget than was available for the PACITA demonstration project.

Lesson 3: Different experience of EU countries with evidence-based policy making are a challenge.

An important aspect of the project 'Future Panel on Public Health Genomics' was cooperation between PTA and non-PTA countries. One respect in which these countries may differ is in terms of the extent to which they have experience with evidence-based policy making. In Lithuania, for example, which is one of the non-PTA partners, links between policy making, on the one hand, and the scientific community or society, on the other hand, are weak. This presents a challenge in general but particularly with respect to long-term policy making on relatively advanced technologies, such as GBITs in health care. Part of that challenge is that some of the non-PTA countries struggle with a lack of basic research and clinical capacities at medical facilities. There may be a clear need in this respect for mutual learning on evidence-based policy making. But it also presents quite a challenge for attaining a clear focus of the policy debate when a participating country is struggling to cover basic needs that need to be met in the short term while the TA debate is focused on long-term visionary goals that involve high-tech such as GBITs. One of the main challenges is the capacity to translate the outcomes of cross-European TA at the national level, taking into account the differences in health-care systems in Europe, technological developments, and financial investments being made into research.

One way of meeting that challenge would be to discuss the potential introduction of GBITs in the context of the sustainability of a diversity of health-care systems in different countries. In other words, for a more relevant and significant impact, cross-European TA should have a clear aim of having a European, national and local integration of results. In the case of the Future Panel on Public Health Genomics, it would thus have been important for small studies to be produced, where the main conclusions of the activity would be analysed considering different national contexts. This would allow the possibility of integrating global and local perspectives, highlighting the main issues of concern, including issues of consensus as well as issues of dissidence. However, this was

DOI: 10.1057/9781137561725.0016

not defined as part of the activity, and therefore, there was no time and budget allocated to it.

Lesson 4: Concerning the role of TA experts, maintain a constructive balance between the role as secretariat and the capacity needed to function as TA specialists

One of the aims of the project was to create learning and mobilization on the potential of expert-based policy making facilitated by TA specialists. In this context, cross-European TA provides unique opportunities to support the development of a collaborative framework between countries with a long experience in doing TA and countries currently initiating TA activities. In practice, however, and mainly due to time and budget constraints, the TA experts involved in the PACITA Future Panel project had to function predominantly as the secretariat of the project. This left insufficient time to properly exchange experiences and expertise between the PTA and non-PTA partners when bringing together the rich and diverse results from the expert working groups in a systematic, constructive and policy-relevant way. One of the ways in which this could be countered would be to more directly involve experienced TA experts from PTA countries in the EWG activities that were led by the non-PTA countries. The fourth lesson learned from the Future Panel on Public Health Genomics, then, is that concerning the role of TA experts, a constructive balance must be maintained between the role as secretariat and the capacity needed to properly function as TA specialists. This lesson also underlines the crucial importance of TA capacity building in non-PTA countries.

Notes

1 These were the aims of the project as specified beforehand (the 'theory'). At several points, there were (small) differences between theory and practice. See A. Krom and D. Stemerding (2014).

2 See http://web.eduskunta.fi/Resource.phx/parliament/committees/future.htx?lng=en.

3 Not long after the start of the project work package (WP), leader IST (Institute Society and Technology) from Belgium was discontinued. The Rathenau Instituut, not previously involved in this WP, took over the role of WP leader.

4 See Expert Working Groups on Public Health Genomics (2013).

DOI: 10.1057/9781137561725.0016

5   See D. Stemerding and A. Krom (eds) (2013).
6   See D. Stemerding and A. Krom (2014).
7   'Why should policy-makers care about public health genomics? Towards a policy agenda' (Paris, 9 June), https://www.eshg.org/satmeetings2013.0.html.
8   In the year 2000, the method was used in a project on the ageing population and in 2005 in a project on Denmark's future energy system. See e.g. Hennen et al. (2004).
9   For a more elaborate comparison, see A. Krom and D. Stemerding (2014).
10 Early on in the project the relative low number of Future Panel members was identified as a potential risk to the project. Subsequently, extensive attempts were made to further expand the panel.

7

The Future of Ageing – Stakeholder Involvement on the Future of Care

Marianne Barland, Pierre Delvenne and Benedikt Rosskamp

▶

Abstract: *Barland et al. describe an example project showcasing the strengths of technology assessment methodology in structuring stakeholder dialogues in a cross-European context. The authors provide an in-depth account of the method design choices made and their underlying rationale. Beyond the buzzword, well-structured and transparent stakeholder dialogue can help to balance difficult issues of policy priority – in this case by balancing the contributions of technological innovation against social reorganization as a means of securing sustainable future health-care service for senior citizens. The article shows the added value of multi-site dialogues based in national debates but linked to the European policy development process.*

Klüver, Lars, Rasmus Øjvind Nielsen, and Marie Louise Jørgensen, eds. *Policy-Oriented Technology Assessment Across Europe: Expanding Capacities*. Basingstoke: Palgrave Macmillan, 2016. DOI: 10.1057/9781137561725.0017.

Figuring out how we can cope with ageing societies is one of the grand challenges identified in the Lund Declaration. The demographic composition of the world is changing, and projections show that in the next 35 years the number of people over 60 years will double, while those aged 80 or older will quadruple. At the same time, the available workforce in the care sector will decrease to a point where the need for care will surpass the available resources. This development challenges existing health-care systems in Europe, and in order to have a sustainable system in the future, one needs to rethink policies related to health care.

The European Commission's 'Digital Agenda for Europe' pointed to technology as part of the solution for addressing the challenges raised by ageing society. The strategy states that new information and communication technology (ICT) capabilities could support ageing citizens, revolutionize health care and provide better public service. But barring the way to any easy technological fix are critical issues, which must be tackled to ensure a sustainable health-care system. Technology will likely be an integral part of such a system, but there will also be a need for substantial social and organizational change to reorganize health-care services in Europe.

To illustrate the value of stakeholder dialogues structured through TA methodology, PACITA organized a cross-European assessment experiment aimed at investigating how technological innovation along with social reorganization could contribute to creating sustainable health-care services for European seniors in the different societal situations of member states.

The project's goal was twofold: (1) to identify opportunities, challenges and barriers as well as policy options for the use of technology in the health-care sector and (2) to train and exchange knowledge on the method of scenario workshops among the project partners and, hence, to increase the national knowledge base for policy making. The result of the project was a series of policy options and recommendations.

Framing the issue of technology and policy in Europe

How or if technology is implemented in the care sector varies greatly among the European countries represented in PACITA,[1] alongside a varied approach from policy makers. In order to map the terrain, the first tasks of the PACITA project on ageing societies were therefore to produce a policy status overview (Fitzgerald, 2014), presenting and comparing the different strategies put forward by policy makers in

DOI: 10.1057/9781137561725.0017

country. In the same way, a technology overview (Meidert and Becker, 2013) was made in order to map the technologies that are used in the European care sector today and to anticipate which technologies may play a role in the future of (health) care for senior citizens.

The technology overview showed that a variety of devices and technology are used in European health-care services today. However, implementation varies from country to country, and the range of technologies is increasing as their market potential is increasingly recognized by developers and investors. Most of the technology, which has already been implemented, belongs to what we may call 'first-generation telecare', such as alarm buttons and sensors. Some countries have already started using more complex technology, which includes the measurement of vital signs or two-way digital communication between patient and doctor to reduce the need for home visits or hospital appointments.

The variation of technologies is reflected at the policy level. Although all countries are facing the same challenges, they respond in quite different ways. Analysis of policy documents from the different countries involved in PACITA shows that the use of technology in care is starting to be recognized in some countries. However, there are large national differences in the way that it is interpreted as well as the perceived level of urgency in designing, addressing and implementing such policies. The analysis of policy documents also shows that there are a number of definitions used to describe telecare and home-based telemedicine. The differences are not only between countries but also within countries – for example, between official governmental reports and national stakeholders.

Technological developments are always difficult to predict, but the technology overview highlights some trends that probably will influence the distribution and implementation of technology in the health-care sector. Among these trends are smartphone and mobile solutions that would enable easier data collection and communication. Together with an increasing use of monitoring devices, digital assistants and a wide selection of apps, mobile health may become a reality in the near future. Data collection and big data analysis will increase and can be used for prediction and preventive work.

Just as important as technological development is the development of regional, national and European policies that address the various ways in which technologies could be integrated in health-care systems. Whether health authorities choose to encourage implementation or to stay passive will strongly affect future use. Private actors and industry will also play

DOI: 10.1057/9781137561725.0017

an important role as the potential of a flourishing market for health-care technology will affect policy making all over Europe. One of the overall conclusions reached in this mapping exercise is thus that long-term policies and strategies will be necessary in order to implement technology in a productive and responsible way.

Engaging stakeholders in policy discussions

There will always be actors that are affected positively or negatively by research, technological development and policy decisions. But often, actors that have a stake in the issues are not automatically consulted or included in the decision-making process, even though they are the ones that will live with the consequences of these decisions. This produces a risk that inappropriate technology may be developed or ineffective policy implemented. In order to avoid this situation, the PACITA project on ageing societies aimed at involving a diverse group of stakeholders to open the discussion to a variety of voices, different kinds of knowledge, perspectives, values and dilemmas.

The underlying argument that supports stakeholder involvement is that it can lead to better-informed policy decisions and more critical discussions about the topic at hand. Typical policy consultations often involve homogenous groups of experts that think along the same lines. Such homogeneity of opinion can weaken the democratic aspect of policy making because the discussion often will evolve around a limited view of the topic. Involving a broader and more balanced spectrum of actors makes the process more diverse and enables the creation of more multi-dimensional and resilient solutions. Additionally, when the concerned actors are included in the process, it can lead to an easier implementation of policy decisions as the involvement facilitates a stronger ownership of the decision-making process among the stakeholders, therefore allowing more robust decisions to be made.

A broadly recruited, heterogeneous group of stakeholders will have very different backgrounds and experiences with a given topic. We therefore developed future-oriented scenarios to give the stakeholders a common starting point for discussion. Using the scenario workshop method, the stakeholders engaged in forward-looking discussions and identified policy options on a given topic. The purpose of the scenarios is to make the participants more conscious of future developments and

DOI: 10.1057/9781137561725.0017

choices related to technology in society and to inspire critical reflection. Through such discussions, stakeholders may contribute to the development and identification of new visions and policy options based on their first-hand experience with the topic at hand.

Creating scenarios for the future of ageing and new technology

Society and policy makers are faced with many collective choices, and the latter need to handle sometimes conflicting priorities when developing their policies. The outcome and the implications of their choices may be difficult to anticipate. Our scenarios on ageing did not try to predict the future and did not purport to encompass all aspects of a possible future. Instead, they presented sharply distinct alternative futures that one might expect to arise from discrete policy choices, highlighting the challenges, dilemmas and conflicts that could occur in order to spur discussion.

It is a challenge to write up scenarios that are considered relevant for a broad group of countries and regions because of how diverse the reality of health-care systems and use of technology are. Immigration, distribution of technology and digital literacy are generally perceived

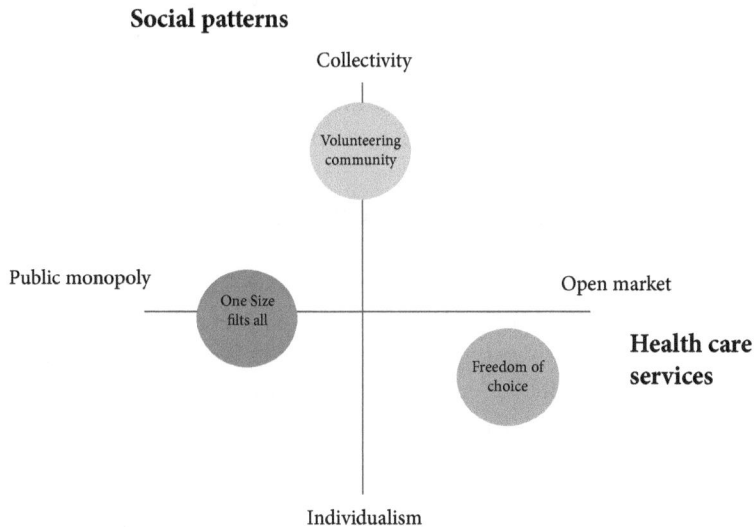

FIGURE 7.1   *The PACITA scenarios for the future of ageing*

DOI: 10.1057/9781137561725.0017

TABLE 7.1   *Content of scenarios on the future of ageing*

The PACITA scenarios on the future of ageing
One size fits all is based on the assumption of lack of labour in the future, and it describes a large-scale governmental initiative that uses technologies to make people more self-reliant. Everyone in need of care is offered a standard 'care kit' that consists of different assistive technologies. Seniors are encouraged to live at home as long as possible.
Freedom of choice is based on a new political system where incentives for care go directly to the user. This scenario furthermore describes a society where you can buy a great variety of health-care services and technology from the open market. Everyone in need of care is entitled to incentives and financial support depending on their individual health condition.
Volunteering community is based on utilizing volunteers as the key resource for the community and for each other. This community could include the senior citizens themselves, their relatives, organizations, neighbours, school kids and so on. The authorities' main responsibility is to mobilize the coordination of the volunteers.

very differently in different countries. Therefore, instead of dealing with concrete technological solutions, we structured ours along two axes of 'social patterns' and 'organization of health-care services'. This more generic approach ensured that all countries could recognize parts of their own reality, but at the same time the scenarios told stories that stirred debate among the stakeholders.

The three scenarios, 'One size fits all', 'Freedom of choice' and 'Volunteering community' describe futures where health-care services are organized and financed in different ways and where health-care service may be affected by increased government control, a stronger private sector or a better organized volunteer community.

Designing national scenario workshops

The main activity in the PACITA project on ageing was ten national scenario workshops organized by the project partners. They all followed the same method:[2] critiquing, discussing and giving feedback on the three scenarios, and in the end formulating visions and recommendations for policy makers. The participants at the workshops were broad groups of stakeholders from academia, the health-care sector, policy makers, public administration, industry and senior organizations.

The results from the workshops were collected in national reports that describe the response to the scenarios and the future recommendations.[3]

DOI: 10.1057/9781137561725.0017

While all countries agreed that there is potential in using technology in the health-care sector, several differences became obvious when it came to describing possible barriers and challenges related to implementation and use. These national peculiarities reflected cultural and social aspects in the respective countries and regions and also reflected to what degree the debate about technology and ageing had been prominent or not. In this way, the differences across countries reflected different values and worldviews with regard to the use of technology in health care and social innovations. In many countries, there were no established arenas beforehand where stakeholders could come together and discuss current and future policy developments. In this way, our experiment was very successful in terms of facilitating dialogue and knowledge exchange between stakeholders that were otherwise unconnected.

Recommendations for future sustainable health-care services

The policy report is structured by five policy issues that were recognized as particularly important at the national workshops, with related policy options and recommendations (summarized in Table 7.2 below).

Technology is considered an important element in future health care by many actors, such as the EU and national or regional authorities all over Europe. The stakeholders involved in the PACITA project support this, but they stressed the importance of broadening the debate and to also look at social and organizational innovation.

Broadening the knowledge base for policy making

Societal challenges that involve new technology can often be perceived as complex and difficult to grasp. The experience from the PACITA project on ageing clearly shows that involving a broad group of stakeholders in discussions can help identifying opportunities, challenges and barriers related to the future of health care and the implementation of new technology. The stakeholders' hands-on knowledge and diverse areas of expertise provided important insights that would not necessarily have been identified by the homogenous expert groups traditionally involved in policy-making processes.

DOI: 10.1057/9781137561725.0017

TABLE 7.2   *Policy recommendations produced by participating stakeholders*

Policy issues	Policy options and recommendations
Support individual needs, self-determination and autonomy	• Enable seniors to live independently and securely at home. • Promote informed decisions. • Improve ageing literacy: prepare seniors for ageing well. • Introduce a system for assessing individual needs.
Provide basic care for everyone	• Develop long-term strategies that responsibly introduce technology and ensure basic care. • Create arenas and networks for knowledge exchange. • Introduce means for prevention of unacceptable consequences, such as loneliness and isolation.
Participation in society and voluntary work	• Establish a mentality and culture for volunteering. • Define tasks and establish trust for participation in care. • Provide incentives for volunteering. • Mobilize senior citizens as active contributors.
Public-private collaboration	• Stimulate service innovation, research and development of telecare and telehealth. • Define infrastructural means and standards. • Require universal design in all services and products. • Stimulate and ensure user-participation in R&I.
Organization, regulation and education	• Protect privacy. • Include technology in education and training of health-care personnel. • Open up for new roles in the health sector. • Focus on dialogue and transparency.

Involvement of carefully selected diverse stakeholder groups is also a way to make policy decisions more democratic, robust and socially acceptable. Involving relevant stakeholders in the process can give them ownership of the process and increases the chances for both adapted policy prescriptions and the development of relevant products actually meeting users' needs. This in turn can make implementation processes easier.

Cross-European stakeholder involvement

The method of scenario workshops has until now mainly been used in national contexts. Using the method in a cross-European manner proved challenging to some degree, but it was also beneficial to the project results and the embedded potential of the method.

In the preparation of the scenarios, it proved challenging to write scenarios that were both general enough to feel relevant for all participating countries and at the same time specific enough to provoke discussion.

DOI: 10.1057/9781137561725.0017

Scenarios that are too general would not have contributed to the desired discussion, while making them too specific would have made it difficult to relate to the range of ethical and social dilemmas to be dealt with. But the cross-European approach proved to give significant added value compared with the more common alternative, which is a series of isolated, national debates taken without much synchronicity. The scenarios created discussion that had the same starting point but that moved in different directions based on national differences in experience, organization and financing of health-care services and national/regional culture, policy preferences and worldviews. The national reports describe dilemmas, barriers and solutions that are grounded in a specific national or regional context but that are highly relevant for policy makers all over Europe.

Realizing that all countries face the same challenge, learning from each other, exchanging experiences and identifying European examples of best practices are starting points for the future of knowledge-based policy making within and across Europe. The method of scenario workshops proved suitable to a cross-European context, and the format of separate national activities that were linked by taking the scenarios as a common starting point for discussion created a common frame for the dialogues which ensured the comparability of the results that were collected at the regional or national level. The PACITA workshops produced important insights for national and regional, as well as European, policy making. But it also highlighted the importance of independent and diverse policy advice, an opinion that was emphasized by all the involved participants. The coming together of stakeholders facilitates not only a knowledge exchange but also knowledge production for the future.

Notes

1   The involved partners represented Austria, Bulgaria, Catalonia (Spain), the Czech Republic, Denmark, Hungary, Ireland, Norway, Switzerland and Wallonia (Belgium).
2   Barland (2013).
3   Country reports are available at www.pacitaproject.eu.

DOI: 10.1057/9781137561725.0017

8

Europe Wide Views on Sustainable Consumption

Marie-Louise Jørgensen, Ventseslav Kozarev and Kathrine Lindegaard Juul

Abstract: *Jørgensen, Kozarev and Lindegaard Juul lay out the rationale and methodology for a multi-site citizen participation exercise carried out within the larger framework of the PACITA project. The exercise gathered more than 1,000 citizens at parallel citizens' summits in eleven European countries, exemplifying the practicability of orchestrating public engagement in connected national arenas across Europe. The authors argue that not only did the events themselves provide comparable samples of informed and deliberated opinions, but also the cross-national collaboration to prepare the events, which involved both central stakeholders and policymakers, served as a vehicle for consensus building among these actors. Based on the response of participants and political recipients, a call is made for further capacity building for cross-European citizen participation.*

Klüver, Lars, Rasmus Øjvind Nielsen, and Marie Louise Jørgensen, eds. *Policy-Oriented Technology Assessment Across Europe: Expanding Capacities.* Basingstoke: Palgrave Macmillan, 2016. DOI: 10.1057/9781137561725.0018.

DOI: 10.1057/9781137561725.0018

The infamous democratic deficit of European institutions has spurred a range of different initiatives that aim to close down the persisting gap between decision makers and citizens. Once a buzzword, public engagement has become a staple of European policy discourse on account of this remaining deficit. By way of realizing the potential of public engagement, procedures have been developed with and in some cases embedded in institutional procedures. But recent cases show that great dissatisfaction among citizens remains with regard to their ability to influence policy.

One promising avenue of development is that of deliberative forms of citizens' engagement at relatively early stages of European policy formation. Such formats have been tested on several occasions by European research projects (CIVISTI, VOICES and others; see also Olsen and Trenz, 2010) and show great promise. These projects have shown the ability of deliberative processes to qualify citizens' opinions based on information and mutual learning as well as to establish through dialogue a democratic rather than merely private mind-set among citizens. This means that while such 'mini-publics' are rarely representative in a statistical sense, they manage nevertheless to give a trustworthy picture of the differences of opinion that may emerge through public debates on policy matters. Furthermore, these experiments have thoroughly debunked the myth that citizens will not be able to grasp the complexity of policy matters. The opposite in fact seems true: citizens quickly home in on the most crucial issues once the knowledge base that is available to decision makers is presented to them.

One reservation remains, however, that prevents Europe from wholeheartedly embracing deliberative public engagement, namely the concern whether citizens are in fact able to adopt a 'European perspective' without the intervention of overly costly procedures of lingual and cultural translation. To address this reservation, the third example project of PACITA adapted a multi-site citizens' participation method developed in the TA community. We wanted to show that the dichotomy between one European policy and several national polities is a false one: national publics are already 'de facto' cosmopolitan publics (Beck and Grande, 2007), and with regard to issues of systemic risks shared across border, coordinating public engagement across European member states in fact produces a genuinely European *vox populi*.

On 25 October 2014, more than 1,000 ordinary citizens participated in this cross-national citizen consultation entitled Europe Wide Views on Sustainable Consumption.[1] The day-long event took place simultaneously

DOI: 10.1057/9781137561725.0018

in eleven EU member states (see below). The substantial aim of the consultation was to bring the reflected views of citizens to policy makers and thus influence concrete policies in the years ahead. Sustainable consumption is one of the grand challenges faced by European society, and one in which the range of policy options is closely linked to public opinion. And policy makers generally hesitate to consider policies aimed at private consumption for fear of intervening too much in the private sphere. With this consultation, we wanted to restructure the debate on policies on sustainable consumption by allowing citizens to redraw from their own perspectives the line between acceptable and intrusive interventions in private consumption patterns. As we shall see below, this public engagement exercise became a process through which not only citizens, but also supporting stakeholders and policy makers came to revisit basic policy assumptions – precisely from a European perspective.

Background

The European citizens' consultation on sustainable consumption was based on a previous method design developed by a core of TA partners, namely the World Wide Views method. This method was originally designed to provide a platform for citizen participation in the UN COP negotiations on climate and biodiversity (Rask, Worthington and Lammi, 2012), but with a few modifications, it proved to be fully adaptable to the European context, producing the 'Europe Wide Views' (EWViews) approach. The method combines simultaneous national face-to-face citizen consultations with a web-based transnational comparison of national results.[2] At each national site, roughly 100 citizens deliberated in small groups on the basis of the same information material and voted anonymously on the same questions which made it possible to make transnational comparisons.

The issue of European policy development for sustainable consumption presents four characteristics, which makes the EWViews method particularly appropriate. First, patterns of production and consumption are intrinsically part of every citizen's daily life, and policies to affect these patterns therefore affect citizens directly. This is the basic criterion for situations in which citizens' participation should be considered a right. Second, the issue is one in which there is knowledge that concerns patterns and options readily available and relatively uncontroversial.

DOI: 10.1057/9781137561725.0018

This means that informing citizens thoroughly and correctly prior to the consultation is possible and that deliberation can start from a platform of evidence. Third, sustainable consumption is an issue field in which political action is necessary at both the European and the member-state level. Market failures produce waiting games in which political intervention at multiple levels of governance is needed to create forward momentum. And lastly, sustainable consumption is an area in which choosing between policy options is an obviously normative, rather than merely technical, issue. The complex interdependencies involved in changing patterns of production and consumption mean that policy choices will have deep ethical, social and distributional effects. This makes the voices of diverse groups of citizens highly relevant since their input will likely foreshadow the reactions of the public at large.

Throughout the process of designing, organizing and carrying out the citizen consultation, politicians, policy makers and stakeholders have continuously been involved in identifying issues for deliberation and balancing sources of knowledge for the information material that was to be distributed to participating citizens. The process was thus supported by MPs, MEPs, Commission staff, NGOs with green and consumer agendas, researchers in the various fields, and interest organization representatives in retail and industry. The immediate purpose of this extensive pre-consultation involvement has been to ensure the direct policy relevance and overall soundness of the citizen consultations and their outputs. But the preparation process in itself has also served as a vehicle of informal dialogue across sectors and has contributed in many small ways to the formation of a common understanding and a common sense of urgency among diverse stakeholder groups. The willingness of politicians and policy-makers to open many of the meetings showed the political interest, which this process generated. The expressed interest of these end users of the citizen consultation made it clear to the participating citizens that the consultation was in fact much more than an academic exercise.

Consultation results

During the citizen consultation, data was collected in two ways. First, at the end of each thematic session, the citizens voted on a set of questions related to the strategies which they had touched upon in their

DOI: 10.1057/9781137561725.0018

TABLE 8.1   *Europe Wide Views in numbers*

Participating countries	11
	Austria, Bulgaria, Catalonia (Spain), the Czech Republic, Denmark, Hungary, Ireland, Lithuania, the Netherlands, Portugal and Wallonia (Belgium)
Participating citizens	1035

deliberations. Second, at randomly selected tables, minute takers reported the views which citizens presented during deliberation.[3]

Generally, the outcomes of the consultation show that the citizens of Europe Wide Views accept the possibility of policy measures aimed at private consumption. Actually, they are strongly in favour of policy makers' taking ambitious steps in order to encourage more sustainable consumption in society. But it's not only policy makers who should take action; citizens also want to be involved in the process of striving towards a higher degree of sustainability in consumption.

Based on a thorough analysis of the quantitative as well as qualitative data, the EWViews partners have agreed on nine policy recommendations. Eight of the recommendations are directly linked to the citizens' views on how policy makers should act in order to achieve more sustainable consumption, while the last one has to do with the future use of citizen engagement in the EU. The nine policy recommendations are presented below in a random order:

- ▸ Set an ambitious European agenda to achieve more sustainable consumption.
- ▸ Perceive citizens as collaborators in striving towards sustainable consumption.
- ▸ Do not leave sustainable consumption solely to the market.
- ▸ Make sustainable consumption cheap and easy.
- ▸ Use financial policy instruments to foster sustainable consumption.
- ▸ Provide better eco-efficient alternatives to conventional car transport.
- ▸ Ensure longer durability of products.
- ▸ Raise awareness and educate citizens on how to consume sustainably.
- ▸ Engage European citizens in dialogue processes in the future.

The recommendations can be studied in greater detail in the policy report.[4]

DOI: 10.1057/9781137561725.0018

Consulting citizens across Europe:
a double question of trust and capacity

As already mentioned, the overall aim of the EWViews experiment went beyond the production of input for the concrete case of European sustainable consumption policy. The exercise was meant also to help build trust in such exercises in general and to spark capacity building among practitioners in the different European member states. The motivation has to do with the state-citizen interaction in Europe. The participation of citizens in policy- and decision-making is increasingly seen as a necessary component of modern democratic societies. Still, EU member states differ in motivations for engagement, in traditions for doing so, in the degree of interest among policy makers and in the perceived legitimacy of such exercises at the policy level. Thus, even if public engagement is a commonly hailed value across Europe, participation exercises do not always succeed in building social trust. This poses a challenge to organizers and champions of participatory processes. Designing successful citizens' participation processes requires thorough and transparent preparation, continuous communication, and mechanisms for follow-up monitoring and control.

Countries handle this challenge very differently. In some countries, public engagement has traditionally been strong and both policy makers and decision makers have frequently based decisions informed by citizens' consultation processes. A few, such as Austria, have frequently relied on referenda, rather than on separate institutions, to encourage the public's involvement in making the decisions themselves. In others, such as Denmark, public engagement traditions have been embedded in the way that specific public institutions are designed, and these traditions are evident in their missions and mandates. Such institutions have been successful in bridging scientific expertise, public deliberations and public opinions and in raising awareness of pending societal challenges, thus contributing to an enhanced policy process on complex and controversial issues.

As a rule, however, in countries without well-organized civil societies and where a closed political culture persists, citizens are only sporadically involved in isolated events and participation is dominated by conflicting reactions rather than proactive dialogue with stakeholders. In these more closed decision-making traditions, decision makers rarely rely on wider public input or simply mirror the demands of disorganized, anonymous

DOI: 10.1057/9781137561725.0018

publics, without real dialogue, analysis or attention to possible impacts. Regrettably, this often translates into the feeling that citizens are being neglected by decision makers and are generally not welcome in the decision-making processes.[5] This is where the build-up of trust in open deliberative processes through concrete experiences is most important and where the hands-on training of practitioners may provide the most value.

For Europe at large, even though traditions and situations vary among countries, seeking larger-scale citizens' involvement with issues that are highly controversial and often not fully understood by decision makers might help reduce complexity and at least help elaborate policy options that can be pursued with a realistic expectation of public acceptance. Organizing such exercises in a manner which coordinates national dialogues to form a European citizens' forum could be viewed as a necessary 'soft' reform of European institutional interaction and a step towards reducing the democratic deficit of the EU.

Lessons learned from EWViews

The consultation was successful across the countries that participated. Participating citizens demonstrated a high degree of support for deliberation and involvement in consulting decision makers. A large majority reported that they would like to see more consultations like the Europe Wide Views in the future, and they expressed that they would also take part in them if they received an invitation. These sentiments were echoed across Europe.

What is of special interest to the agenda of expanding TA is that in those countries without established TA institutions, the national events managed to stir up debate and create a focus on citizen engagement. Furthermore, the perceived legitimacy of the events was high due to the transparent process of consultation, which was perceived as trustworthy by participants and recipients alike. Most of the participating citizens reported that they for once felt included, and they were therefore pleased to express their opinion, as they knew it would be considered by policy makers.[6]

The EWViews method proved to travel well. Citizens' engagement in national deliberations was very lively in all countries. In part, this was due to the presence of skilful moderators, but to a much higher extent

DOI: 10.1057/9781137561725.0018

to the fact that the participating citizens felt that they had a voice to be heard. They could, and often did, relate to their own experiences, and they provided numerous examples to support their arguments. All deliberations were markedly based on dialogue and respect, which contributed to the sense of accomplishment at the end of the day.

In terms of preparation, the greatest challenge turned out to be the recruitment of participants. Citizens in some countries remain very reluctant to share their opinions in public. Even among those who agreed to participate, some were hesitant at the beginning. The moderators, however, were prepared for such a challenge and helped create a very positive atmosphere at each table, helping citizens overcome their hesitation. Over time, the best remedy for this hesitancy will likely be further experiments that expose growing numbers of citizens to the participation experience, which would help to increase capacities and create a virtuous circle of growing trust among citizens in such processes.

Future perspectives and conclusions

The citizens' evaluation demonstrated that the consultation was successful. The overwhelming support for engaging citizens more in decision-making processes was equally present in countries with extensive as well as little experience with citizen-participation processes. A Walloon citizen expressed his support for more citizen engagement in the EU, in the following way:

> Envision more frequent consultations of active citizens, of people wanting to take part in debates. Citizen dynamics such as this summit should be systematized.

Furthermore, the citizen consultation was also a success from a public-policy point of view. It has produced a set of very clear policy recommendations on how citizens think that policy makers should act in order to achieve a higher degree of sustainability in consumption. We hope that policy makers will make use of the unique insights into the views of ordinary citizens and will carefully consider them when formulating future policies that relate to sustainable consumption.

Additionally, the fact that the citizen consultation took place *simultaneously* in the eleven countries helped to give participants a sense of

DOI: 10.1057/9781137561725.0018

being part of something bigger, that went far beyond the walls of their respective national meetings: a truly European event. Therefore, Europe Wide Views is also a way to emotionally minimize the distance between citizens across EU member states and hereby strengthen the European community.

To harvest these fruits, a more systematic use of similar methods for participation in the future could help build capacities and pave the way for both the formal and the informal acceptance of citizens' engagement within the governance institutions of Europe and its member states. Such systematic development would provide evidently added value from a European perspective.

Notes

1   National holidays meant that Czech and Hungarian meetings were held one week earlier.
2   For more information, visit http://www.wwviews.org/.
3   Minutes were taken in national languages and qualitative reports translated to English.
4   Policy report with results comparison functionality are available at www.citizenconsultation.pacitaproject.eu.
5   An opinion strongly expressed in Bulgaria, the Czech Republic and Hungary during the national EWViews consultation on 25 October 2014.
6   This was particularly evident in Hungary and Bulgaria.

Part III
Building Capacities for Cross-European TA

▶

DOI: 10.1057/9781137561725.0019

OPEN

9

Making Technology Assessment Accessible to New Players

*Pierre Delvenne, Benedikt Rosskamp,
Ciara Fitzgerald and Frédéric Adam*

Abstract: *Delvenne et al. present theoretical considerations about the pedagogy of technology assessment (TA) in general and the summer school format in particular, which was chosen as a platform for teaching TA in the PACITA project. The PACITA summer school programme was designed to encourage the uptake and use of TA rationale and methods by various types of professionals involved in science, technology or innovation policy. The recruitment strategies, the format of the presentations, and so on of the two summer schools are presented. The authors argue that as the 'responsible innovation' agenda gains traction among policy makers, societal actors and academics, education initiatives such as the TA summer school can have an important role to play in shaping understandings of this new form of governance.*

Klüver, Lars, Rasmus Øjvind Nielsen, and Marie Louise Jørgensen, eds. *Policy-Oriented Technology Assessment Across Europe: Expanding Capacities.* Basingstoke: Palgrave Macmillan, 2016. DOI: 10.1057/9781137561725.0020.

DOI: 10.1057/9781137561725.0020

This chapter reports on the two PACITA summers schools, which were aimed at teaching TA as well as enhancing mutual-learning activities. The first summer school concentrated on 'Renewable Energy Systems' role and use of PTA' and it was held in Liège, Belgium, in June 2012. The second summer school addressed the topic of 'Ageing and Technology' and was held in Cork in June 2014. We describe the rationale and format of the summer school in order to present a comprehensive account of how it introduced TA, both its rational and its methods, to a new audience. We argue that as the responsible innovation agenda continues to gain traction among policy makers, societal actors and academics, education initiatives such as TA summer schools can have an important role to play in the future of the governance of science, technology and innovation.

Background and rationale

Training and learning activities in TA encompass a great variety of approaches, including embedding TA-like courses into engineering and natural scientific curricula or TA practitioners training. In the former case, the objective is to raise students' awareness of social and ethical dimensions relative to technology development and implementation. But in the latter case the objective is to exchange best practices and, by doing so, constituting a community of practitioners and even a scientific (inter)discipline that goes beyond the established community of TA practitioners. However, along these already existing activities, which are organized and implemented in a number of ways in European countries, the PACITA project stressed that in a context in which knowledge-based policy making is increasingly needed, very few TA training activities *directly* target policy makers. This creates two major difficulties. First, a broad set of policy makers and innovation actors from countries where TA institutions are already established, when they are aware of what TA is, might not be conscious that they could use already existing TA knowledge to address the policy-making issues that they are confronted with. Second, in countries where TA practices are not institutionalized as such, policy makers may fail to support the need to further establish such activities, more by lack of knowledge about TA rather than by lack of enthusiasm. This calls for a need to provide them with convincing evidence that TA knowledge is of valuable potential for their daily work.

DOI: 10.1057/9781137561725.0020

In what follows, we argue that the further development of training activities such as TA summer schools is a relevant tool for doing so.

In PACITA, the rationale of TA summer schools was to broadly consider potential *users of TA knowledge*, such as policy makers, civil society organizations, scientists, science communicators and journalists, as well as civil servants, and to sensitize them to the role and added value of TA to their working practices and organizations' objectives. In line with PACITA's aim to expand the TA landscape in European countries which do not count institutionalized TA bodies, summer schools explicitly (though not exclusively) targeted new players in such countries – for example, Belgium, Lithuania, Bulgaria, Portugal, Ireland, Hungary or the Czech Republic. Furthermore, the summer schools also engaged participants from countries with established TA institutions who do not always recognize their TA activities because they believe they do not appear as the main addressee of TA activities. Lastly, the summer schools offered an opportunity to open up and sensitize TA and knowledge-based policy making beyond the fifteen countries and regions represented in the PACITA consortium. The events attracted participants from EU-28, Africa, Australia, South-America and Asia.

Overview of the two summer schools

The two summer schools' topics were centred on two 'grand challenges for Europe', particularly suitable to technology assessment approaches and methods. In Liège 2012, the topic was renewable energy systems, while in Cork 2012, the summer school there focused on ageing societies and new technologies. The complexity of these grand challenges and the great transitions that they necessitate appeared to be adequate backgrounds to call for new modes of interaction and exchange with and among 'new players' in technology assessment.

The first summer school[1] was organized at the University of Liège, Belgium (25–28 June 2012). As a transnational concern and growing grand challenge for policy, economy and society worldwide, the topic of 'renewable energy systems' was chosen as an entry point for learning about TA. This challenge refers to the interplay of actors, technologies, policies, worldviews and institutions engaged in the field of energy debates, policies and production. Technologies play an important role in coping with such issues. At the same time, technologies can also be part of the problem. Participants at

DOI: 10.1057/9781137561725.0020

the summer schools were taught balanced, encompassing approaches and relevant TA methods to address the most pressing energy issues.

The second summer school was organized at the University College Cork, Ireland (17–20 June 2014). The topic chosen was 'challenges and opportunities of the ageing society: exploring the role of technology'. The event consisted of training sessions, practical exercises, mutual reflection, and networking. Figuring out how to cope with ageing societies is one of the grand challenges pointed out in the Lund Declaration, and health-care technologies can be increasingly important for society to offer health and care services at a quantity and quality that mirrors the expectations of the European population. The summer school participants debated how best we can use new technology in care services and what type of policy options policy makers are faced with.

Summer school format

Summer schools were a combination of lectures and interactive workshops. Lectures combined elements of the different phases of a TA project (problem definition and research design, methodology, communication and impact) with concrete examples or applications to the issue at stake. After each lecture, during the workshops the participants would have the chance to relate what they had learned in hands-on, problem-driven simulation and role-play exercises. The workshops' objective was to produce a coherent draft for a TA project. A facilitator helped participants with a 'script' that included minimal contextual information (such as the context in which a TA project was needed or the explicit demand from a politician's commissioning a study) and suggestions for sub-tasks (identifying the needed knowledge base, mapping relevant stakeholders, listing technological options, scrutinizing social issues as well as more practical tasks such as project management and communication).

Participants were split into two groups, and they were assigned different roles within the workshops, as happens in real TA institutions (e.g. researchers, project managers and communication officers). Before they started working, each group was given different variables such as the addresses of the project, the framing of the issue, the available budget, the timeframe for decisions to be made, the technologies involved, the existing expertise, the mapping of stakeholders or the socio-political context. Both groups were also given different assignments. This could for instance

DOI: 10.1057/9781137561725.0020

be a study that originated from a member of European Parliament's demand or from setting up a new project on a city level to then present it to TA's addressees. This resulted in the two groups presenting contrasting approaches, project management's choices and expected results. To finalize the training, the groups presented their work to each other in order to exemplify the diversity of possible TA approaches on a complex issue.

Main results

The summer schools can be considered as a first step in the construction and consolidation of an international *TA community extended beyond the TA practitioners themselves*. Numerous participants have kept in touch and established collaborations. Furthermore, once participants were introduced to the concept of technology assessment, they also attended other events in the TA community and particularly within the PACITA project, such as the Prague Conference or the practitioners training activities. In addition, the TA simulation exercises facilitated a common understanding and shared interest in TA, thus indirectly strengthening the support base for establishing TA in other European countries. Summer schools also confronted TA practitioners with various ontologies of technology assessment.

Lastly, for participants and TA practitioners alike, summer schools provided a platform for mutual learning, not only about technology and grand challenges but also about the views of various societal actors on TA. This continuous iterative learning approach is especially relevant in the context of expanding the TA landscape, as it helps provide the traditional TA players with a feedback mechanism from the new players who are sensitized to what TA is and what it can deliver.

Future agenda for TA education in the context of 'responsible innovation'

Today, with the discourse of addressing grand challenges (especially in the European Union; cf. Lund Declaration or Horizon 2020), the promises of and strategies for technology are not yet very specific. At the same time, it has become widely acknowledged that governing grand challenges is a complex issue that requires knowledge-based policy-making solutions.

DOI: 10.1057/9781137561725.0020

These evolutions call for recognition of the importance of governance, the broadening of government and the inclusion of more actors in collective choices that involve science and technology. Governance is actually distributed between a number of actors, which some definitions acknowledge: governance can be discussed as the coordination and control of autonomous but interdependent actors either by an external authority or by internal mechanisms of self-regulation or self-control (Mayntz and Scharpf, 1995, Benz, 2007), including *de facto* governance arrangements that emerge and become forceful when institutionalized (Kooiman, 2003). With such a notion of governance, it becomes understandable how the trend of grand challenges impinges on the governance of science, technology and innovation and how anticipating future developments and relating them to policy making has become a crucially important task for technology assessment.

In a first attempt at discussing the anticipatory governance of science and technology, Barben et al. characterized anticipatory governance as evoking a distributed capacity for learning and interaction stimulated into present action by reflection on imagined present and future sociotechnical outcomes (Barben et al., 2008: 993). On these grounds, summer schools can be taken as practical instances of anticipatory governance because they emphasized broadening the community of TA users and enhancing a distributed capacity to frame cutting-edge issues in terms coherent with TA frameworks and tools. An important lesson learned has been that TA knowledge is not produced by one actor in isolation before it is transferred to other actors deemed to use the subsequent insights. Rather, TA knowledge is co-produced by a range of actors who contribute in order to collectively generate knowledge resources, partly already informed by governance issues.

Recently, there has been increasing attention to that idea in connection with policy discourse on the concept of Responsible Research and Innovation (RRI). One influential definition of this concept combines good intentions with anticipation and mates it with attempts at anticipatory governance (Owen, Bessant and Heintz, 2013). In this definition responsibility has a prospective element (it is more than accountability) and 'responsible development' is a multi-actor distributed process. Therefore this type of governance qualifies as anticipatory governance. There are bottom-up dynamics, but at the moment, the policy discourse is most visible. More should be done in order for the policy discourse to be more firmly and systematically entrenched in bottom-up innovative

DOI: 10.1057/9781137561725.0020

practices. Training new practitioners and potential users of TA, like it was done in the summer schools, adds a practical dimension to the debate and contributes to the European strive for ensuring societally responsible research and innovation.

Note

1   See also the article by Pascale Messer in the VolTA magazine: http://volta. pacitaproject.eu/pacita-summer-school-2012/.

DOI: 10.1057/9781137561725.0020

10

Training TA Professionals

Danielle Bütschi, Zoya Damaniova,
Ventseslav Kovarev and Blagovesta Chonkova

Abstract: *Researchers, project managers and communication officers involved in TA projects are faced with a variety of context-dependent challenges which necessitate that TA practitioners constantly reflect upon their practices, innovate and strengthen their skills, making knowledge sharing essential. In the light of this, Bütschi et al. investigate the needs for and possibilities in practitioners' meetings and debates the different needs from established and newcomer TA organizations. The authors convey lessons learned from four PACITA practitioners meetings about principles of knowledge sharing useful for practitioners' training in the future. And they argue for the necessity for TA institutions and their supporters in European policy to use future implementations of similar formats as a way of building human capacities for TA.*

Klüver, Lars, Rasmus Øjvind Nielsen, and Marie Louise Jørgensen, eds. *Policy-Oriented Technology Assessment Across Europe: Expanding Capacities*. Basingstoke: Palgrave Macmillan, 2016. DOI: 10.1057/9781137561725.0021.

In this chapter, we discuss the needs for TA professionals' training, taking into consideration both the needs of established TA organizations, as well as those of organizations trying to develop TA activities in their countries. Based on concrete experiences, we shall draw some conclusions on the contribution that training TA professionals has in strengthening and expanding the TA landscape in Europe.

The attainment of an open, inclusive and transparent governance, as well as evidence-based policy making in Europe, requires the development and further enhancement of capacities for providing insight into the opportunities and consequences related to science and technology, by facilitating democratic processes of debate and awareness building and by formulating policy options in the field of science, technology and innovation (STI). Various organizations in Europe undertake activities that are included in the concept of TA. Yet, TA is still performed by relatively small and mostly nationally/regionally focused institutions, which do not have the needed resources and/or the mandate to make the necessary effort to expand the capacity and use of knowledge-based policy making in Europe. In addition, there is a growing tendency in the field of science and technology to move decision making upwards (from the national to the European level), which entails a common effort and a consolidation of expertise from across Europe in doing European-level TA. Furthermore, considering that in many countries there is no institutionalized approach to doing TA, training professionals from those countries is needed in order to strengthen national capacities for evidence-based policy making. These were among the major motivations to form the PACITA consortium and include TA practitioners' training seminars as an integral part of the work programme of the project.

The PACITA training seminars aimed to stir the communication and mutual learning among TA practitioners. They were designed so that researchers, project managers and communication specialists could learn from each other by sharing their knowledge and best practices. Considering the large variety of TA settings in Europe, the training seminars were conceptualized so that participants who aspire the establishment of TA in their own country could learn about the challenges and solutions related to the different settings of TA institutions; they could thus enhance their understanding of TA approaches and methods and increase their capacities in providing knowledge-based policy

DOI: 10.1057/9781137561725.0021

advice on science- and technology-related issues. For the professionals who work in established TA institutions, the PACITA training seminars offered an opportunity to broaden their practical knowledge as they could become inspired by the work of their colleagues and share best practices.

Shared knowledge for a strong and innovative TA community

The way of doing TA is strongly related to the specific cultural and political environment of a country – as well as to other institutional aspects, such as whether there is a formal link to the parliament, the available funding, its source and so on. This is reflected in the various approaches and methods used within the TA community. This diversity of practices makes technology assessment an innovative and dynamic community, to which many professionals and scientists contribute. But for TA to be more than an experimenting field and for it to become a community that shares a common vision and relies on specific tools, it is important that TA professionals draw on a shared knowledge of what technology assessment is, how it works and what it can achieve. All these aspects are actually covered by extensive literature on technology assessment (see for example Vig and Paschen, 2000, Decker and Ladikas, 2004, Grunwald, 2009 and Enzing et al., 20112), which provides the core elements for the daily practices of TA professionals. However, TA project managers, researchers or communication officers are often confronted with very concrete issues which are not (or are only partially) covered by the literature. What they need is very practical advice related to TA project management: how they should design and frame a concrete project, which methods they should select and how they should implement them, how they should deal with the political and societal environment and how they should communicate their results. For the TA community to further develop and adapt to the ongoing technological and policy changes, it is essential to develop European-wide training platforms, wherein TA professionals will get the opportunity to learn from each other and to work in a systematized and integrative way. This is necessary to ensure a high and uniform level of quality for TA across Europe.

DOI: 10.1057/9781137561725.0021

The PACITA practitioners training seminars

The need for an integrative and systematized training of TA professionals has been recognized some fifteen years ago by the European Parliamentary Technology Assessment (EPTA) network. Since the end of the 1990s, EPTA organizes TA practitioners' meetings once in every two years. Each workshop is hosted and organized by a different EPTA member. Themes address common aspects of TA work, such as determining TA-relevant issues, defining TA projects, communicating TA results, and so on.

The PACITA project continued this tradition by organizing four practitioners' training seminars, which took place between September 2012 and September 2014. Each seminar lasted three days and gathered about 30 TA professionals from all over Europe. The seminars were open to all institutes that perform (or that intend to perform) TA, regardless of whether they are involved in the PACITA project. PACITA covered the costs of the host, as well as travel and accommodation expenses of PACITA partners (others had to pay from their own funds).

The trainings were designed to address the four main stages and the major challenges that project managers face when they run TA projects:

▸ The first essential challenge that TA practitioners have to deal with is the identification and framing of the issue to be addressed. TA projects have to be based on a prior monitoring process of science and technology innovations and of their societal implications; the social and political context has to be clarified as well. During the first training seminar, participants worked on case studies and shared experiences on how they select and define TA-relevant issues.

▸ A second challenge lies in the selection of a relevant method or relevant methods for meeting the project's goals. This issue was addressed during the second training seminar as participants worked through fictive (but reality-inspired) case studies that featured a contentious TA topic and that demonstrated the complex linkages between societal challenges, technology options and policy solutions. Specific application strategies, complementarities of different TA methods, methodological planning and project designs were then explored in greater depth.

▸ During the course of TA projects, various stakeholders need to be involved, which is a challenging task for TA professionals. The third

DOI: 10.1057/9781137561725.0021

training seminar focused on questions: Which actors need to be involved in TA? Why and how are these actors important? What is their role? What are the main challenges for engaging them?

▶ And last but not least, as TA aims at advising policy making on technological and scientific issues, TA practitioners have to communicate the results of their projects. Communication strategies and tools for communicating the results of a TA project were the central theme of the fourth practitioners' meeting.

All the trainings involved intensive group work, plenary presentations and plenary discussions. This proved to be a particularly inspiring experience for newcomers in the TA community, as they could gain insights into the practicalities of doing TA and integrating science and technology into social discourses, public policies and decision making. More experienced TA professionals also could gain practical knowledge for their daily work and extend the professional network they can rely on for future activities. When the participants were asked about the benefits of such trainings, two thirds of them indicated that they had gained new knowledge on TA and half of them indicated that they had learned new TA skills. Most of the participants said that they extended their professional network and found inspiration and new ideas for their work. On average, respondents rated the usefulness of such meetings 5 on a scale from 1 to 6.

Expanding the TA landscape through training

In many countries where no institutionalized approach to TA exists, we can find organizations implementing TA-like activities such as foresight projects and inter- or trans-disciplinary researches or participating in European initiatives that involve the use of technology assessment methods. Yet, in order to be able to lay the groundwork for knowledge-based policy making in these countries, it is important for these organizations to increase their understanding of how TA is done in different political settings so that they can support the process of expanding TA in their own countries.

The PACITA practitioners' training seminars proved to be very helpful in this respect. Interacting with professionals from already established TA institutions and listening to their experiences in TA during the

DOI: 10.1057/9781137561725.0021

training sessions was a great learning opportunity for 'newcomers' in the field. They could get to know the criteria used to select and frame the issue under scrutiny, different approaches for selecting relevant TA methods, the available input and needed outcomes and various other factors. The participants could also learn about when and how to involve stakeholders, civil society and policy makers in the TA processes and how to communicate the achieved results. Some of the major insights in this respect concern the role of actors, which is liable to change over time and over the different project phases; the potential conflict between evidence-based policy making and the political agenda of policy makers; the importance of making the policy cycle transparent to the stakeholders who were involved; and the difficulties in initiating dialogue among the stakeholders and the importance of using appropriate language for communicating with politicians and citizens. In this respect, practitioners' meetings proved to be especially fruitful to those who are looking for national proponents of TA within their own countries and attempting to demonstrate the relevance of TA in their national contexts. Not only could partners from countries with no TA traditions learn first-hand from the experienced partners, but also they could expand their network and thus strengthen the foundation for successfully establishing and implementing TA in their country.

Review and perspectives

When we look back at PACITA TA training seminars (as well at the past EPTA practitioners' meetings), such events bear significance for both established TA institutes and organizations that are developing TA activities in their country or region. However, organizing such trainings implies the availability of funds not only for the organizers but also for the participating organizations. Whereas established institutes may have the resources to organize practitioners' training seminars and finance the participation of their staffers, the situation is more problematic for institutes which have scarce resources. The fact that the European Commission provided funds to the PACITA consortium to organize such a series of events was clearly an advantage, as all member institutes of the consortium could send their staffers regardless of their financial situation. Supporting the organization of training events that help with building specialized and policy-relevant knowledge and skills, such

DOI: 10.1057/9781137561725.0021

as TA, could be prioritized in the European research and innovation programmes. By this, the European Commission will stimulate continuing collaboration among diverse organizational partners and will also include a larger set of practitioners. Not least, however, such a high-level programming commitment will additionally legitimize the application of TA methods in support of policy design and development regarding science, technology and innovation.

For the future, it might also be worthwhile to look for new tools for knowledge transfer that complement the training seminars. Such tools would be important to make the topics presented and discussed during the training seminars accessible to a wide audience of professionals, and also to deepening their knowledge on certain aspects of TA or specific TA methods. In that respect, a series of manuals or best-practice reports could be initiated. New online tools may also be developed.

The issues to be addressed in training, be they in the form of seminars or of written tools, are manifold. The idea of covering the major steps of a TA project in the four PACITA training seminars has been considered by the participants as a meaningful approach. However, participants suggested additional topics of interest, such as determining which are the most pressing issues to which TA could contribute (technology scanning), presenting current TA projects and different TA organizational settings, discussing the specificities of TA project management, exploring possible ways of collaboration between TA institutions and assessing the role of TA contributions for the governance of science and technology. Some participants also suggested integrating better the needs and expectations of the decision makers, who are the end-users of the TA activities. There is obviously a need for TA professionals not only to learn about and share what technology assessment is and how to do it but also to meet with and learn from their addressees. Similarly, the idea of inviting journalists has been raised; their presence would provide an 'insider' perspective on ways to go public or, in some cases, to enable journalists to understand better the communication aspects of a TA project.

The PACITA practitioners' meetings had the particularity of being practice-oriented: concrete TA projects were presented in terms of good practices, and activities were proposed to participants. When ask about this format, three thirds of the participants of the PACITA training seminars wished that future practitioners' trainings would dedicate more time to theoretical aspects of TA or the topic at hand, and more than three quarters would like to have more time for the discussion of case

DOI: 10.1057/9781137561725.0021

studies in terms of best practices. This demand for more theoretical and case study presentations actually calls for complementing the practitioners' meetings with written material that presents theoretical aspects of TA-as-a-practice as well as case studies and best practices in a comprehensive and accessible way. Thus, TA-relevant knowledge would persist and could be utilized in subsequent projects.

DOI: 10.1057/9781137561725.0021

11

Building Community – Or Why We Need an Ongoing Conference Platform for TA

*Constanze Scherz, Lenka Hebáková,
Leonhard Hennen, Tomáš Michalek, Julia Hahn
and Stefanie B. Seitz*

▶

Abstract: *As a background for current outlooks towards
strengthening the technology assessment (TA) community,
Scherz et al. give a historical overview of efforts to establish
international fora for communication among professionals
and researchers in TA. Against this background, the
article conveys experiences from the first two bi-annual
TA conferences, arranged in the context of the PACITA
project. The authors describe experiences of mutual learning
across national boundaries and communicate a renewed
understanding of the necessity for supporting TA capacities at
the national level through professional community building.
Ultimately, Scherz et al. argue that a European TA platform is
necessary for establishing a common language for TA and for
supporting the spread of TA across borders.*

Klüver, Lars, Rasmus Øjvind Nielsen, and Marie Louise
Jørgensen, eds. *Policy-Oriented Technology Assessment
Across Europe: Expanding Capacities.* Basingstoke: Palgrave
Macmillan, 2016. DOI: 10.1057/9781137561725.0022.

Conferences are a promising format to include an extended range of European, national and regional stakeholders – especially with a focus on widening the debate of TA in Europe. Therefore, they are important under several aspects: for scientists from several disciplines in order to discuss inter- and trans-disciplinary approaches and projects as well as for TA researchers to get in contact with their target audiences, such as citizens, policy makers or scientists from other disciplines.

This chapter deals with the question of how conferences can encourage mobilizing stakeholders to establish TA capacities while creating awareness regarding the benefits of cross-European TA throughout Europe. Thus, it reflects on the format of TA conferences as such and gives brief insights into two international conferences, which took place in Prague (2013) and Berlin (2015). Our main argument is that TA can act as a 'knowledge broker' between scientists and policy makers (Riedlinger, 2013). In our experiences, TA and its conferences can provide unique spaces for 'discourse'. Yet at the same time, these discourses need continuity and ongoing activities, which include already established networks as well as new contents, methods and people.

It is in these spaces for discourse that the conceptual basis of TA is reflected upon and further developed. Being a problem-oriented approach, TA needs areas of exchange to enable 'identity-shaping' and adaptation to current challenges. Especially in contexts where its institutionalization is still under development, TA requires formats, which enable mutual learning and critical self-reflection. With recent concepts such as Responsible Research and Innovation emerging, TA has to reflect on how it can contribute and/or offer its wide experiences in various contexts. Further, the format of conferences also offers a useful and inspiring atmosphere for younger researchers and practitioners who are working in the field of TA to present themselves and their questions and to engage in exchange with the wider TA community.

The ambitious goals of the two conferences within the PACITA project were to address the grand transitions and grand challenges that define our societies as a whole. This frame set the scene for presenting and discussing TA research at the conferences and at the same time for offering fruitful spaces of encounter to further strengthen and foster TA as a concept and approach by including all its significant actors (e.g. researchers, practitioners and policy makers). For this, it also seems important to reflect on the experiences already made with international

DOI: 10.1057/9781137561725.0022

TA conferences within the community in order to guarantee a high quality of conferences' input, integrative formats and inspiring topics.

Making it work – the context of the two European TA conferences

As a mobilization and mutual learning project, PACITA aims to bring together established TA institutions and new actors. Consequently, scientific conferences are at the very heart of the project's mission: they intensify the debate on TA and have the potential to expand the landscape of TA in Europe. There is a special focus on the methods and activities in which citizens and policy makers are directly involved in debates and discussions. 'Such "interactive" methodology has proven to be a specific trademark for Technology Assessment and is of special interest today when the focus of research and innovation is turned towards the Grand Challenges of our societies' (Klüver, 2014: 12). Further, conferences provide a platform for scientists with practical experiences as a result of doing TA and for politicians that are addressees of TA research and its results. The two PACITA conferences, held in 2013 and 2015, were the first European TA conferences in more than two decades. In general, the feedback from the conference attendees showed clearly the need for further continuous exchange, networking, discussions and documentation. 'Technology Assessment has shown to be a practice still in the making and continuously expanding its reach and borders, which gives hope for a future with a larger and more branched-out professional community' (Klüver, 2014: 12).

These two major European TA conferences fostered and enhanced the scientific debate about TA as well as the exchange of TA experiences on a European level. The main aim of these and PACITA's ongoing activities is to establish a European network of institutions and persons from the academic world, from scientific policy advice and from policy making. The conferences present an important context for this. With an informative and interactive format, the conferences aimed to bring together several different disciplinary communities. Adopting a broad understanding of what qualifies as 'TA' allowed the conferences to address TA practitioners, academics, scientists, policy-makers, and CSO representatives together. In retrospect, the conferences succeeded in delivering a two benefits

DOI: 10.1057/9781137561725.0022

ways. On the one hand they offered a broad platform for presenting and reflecting on project results, its outcomes and new insights. On the other hand, they helped to set the stage for current and future thinking about TA and its role in tackling the societal challenges ahead.

No future without a past

In order to reflect on the necessity of an ongoing conference platform, it is helpful to have a brief look at the historical development of the TA community in Europe. The major strands of development show that there is a shift from national activities to cross-European and international activities. Also there is an interest in widening the disciplinary community to inter- and trans-disciplinary work. The first meeting of the European TA community under the label of 'European Congresses of Technology Assessment' dates back to October 1982 when the Ministry of the Interior of the Federal Republic of Germany hosted a conference in Bonn that attracted some 60 experts from eleven countries – among them were representatives of the US Office of Technology Assessment. Congresses on TA later held in Amsterdam (1987), Milan (1990) and Copenhagen (1992) contributed significantly to the conceptualization, philosophy as well as institutionalization of TA. These conferences made clear that the European debate on TA took place on several levels – between international groups of scholars, experts, and officials who held a series of meetings during which methods of TA, the utility of its results and the possibilities and problems of institutionalizing TA agencies were discussed.

Another ongoing activity is the institutionalization of networks. During the last ten years, the institutionalization of the German-speaking 'Network Technology Assessment' (NTA) can be seen as a forerunner. Founded in November 2004 in Berlin, NTA aims to identify joint research and advisory responsibilities, to initiate methodological developments, to support the exchange of information and to strengthen the role of technology assessment in science and society. Today, ten years after this first meeting, there have been six scientific NTA conferences, ten annual member meetings and several meetings of the Network's working groups. The primary mission of NTA remains: to provide a platform for information and communication among scientists, experts and practitioners who work in the wide range of TA-relevant topics.[1] The NTA conferences are the central format of exchange among the

DOI: 10.1057/9781137561725.0022

German-speaking TA community. With decades of experience, the three main organizations of the Network for Technology Assessment (NTA) – the Institute of Technology Assessment and Systems Analysis (ITAS) in Karlsruhe, Germany; the Institute of Technology Assessment (ITA) in Vienna, Austria; and the Center for Technology Assessment (TA Swiss) in Berne, Switzerland – also brought their expertise to the PACITA project. Also, other PACITA partners, such as the Danish Board of Technology, the Norwegian Board of Technology, the Advisory Board of the Parliament of Catalonia for Science and Technology and the Rathenau Institute from the Netherlands have worked intensely and enduringly to realize TA in and for parliaments. Together with institutions from Finland, France, Greece, Italy, Sweden and the United Kingdom, they are organized in the European Parliamentarian Technology Assessment Network (EPTA), which was established in 1990 by the president of the European Parliament.[2]

In general, the two PACITA conferences benefitted greatly from these traditions. The conferences of the 1980s and 1990s gave first insights into which topics were relevant for research and policy advice. They also showed how important it is to invite both the scientific community as well as practitioners and policy makers to one and the same event, enabling networking and cooperation on an international level. The EPTA network in particular was and still is exceptionally important to bringing up TA-relevant research topics to national parliaments. For the two PACITA conferences, these contacts are crucial to continuously strengthen the European TA community and to bring together interested researchers, stakeholders and politicians from all over the world. In the days of globalized problems like climate change or world-wide trade networks, this internationalization aspect is of special importance.

Overcoming challenges – making cross-European TA conferences

Generally, doing TA in Europe still remains a challenge. The broad variety of the topics and the positive resonance to the conference show that there was a great necessity to revive the tradition of European TA conferences. It is a substantial gain that TA practitioners and policy makers from countries with established TA practices were able to get involved in discussions with colleagues from countries where TA is still in its beginnings, not only to give advice but also to reflect on their own traditions and established

DOI: 10.1057/9781137561725.0022

TA practices. Besides the national perspectives, cross-European TA must, among other obstacles, face the tension that may arise between the different levels of decision-making structures: European ones versus national and local ones. Which TA topics will be important and popular during the coming years? What can scientists learn from their experiences of working together with stakeholders and politicians?

The two conferences, namely in Prague (2013) and Berlin (2015), clearly showed that there is a strong European TA community interested in joint work and scientific exchange – in spite of sometimes significant differences in the TA approaches that they respectively follow. In Germany, for example, TA institutions work closely with policy makers and politicians. In Denmark, TA institutions strive to fulfil the politicians' needs with a more service-oriented approach. On the other hand, in the Netherlands, there is a certain distance between them. In the so-called TA-emerging countries, technology assessment is yet to be institutionalized. There are many ongoing TA-like activities in countries such as the Czech Republic and Poland – research and development mainly focus on forward-looking studies and methods. But also experiences from beyond Europe are valid contributions. For example, in Japan, as a result of the Fukushima nuclear accident in 2011, the government is trying to recover the lost public trust, by launching an innovative education and research programme that includes TA, which was introduced for the first time in history. These various situations show the challenges and specific situations that TA faces (Michalek et al., 2014). Moreover, spreading the TA community eastwards brings up yet another challenge of finding a 'common language' (Nierling et al., 2013: 105). Due to the fact that TA as such is not institutionalized in the TA-emerging countries, the practices and relevance of such an approach are still being understood differently: 'The processes of institutionalisation of TA infrastructures are always embedded in the understanding of democracy and the role of (national) parliaments' (Nierling et al., 2013: 102).

The PACITA conferences were especially important for TA researchers, in order to get closer to their clients – be it citizens, policy makers or scientists. As David Cope summarizes,

'like any congregation of specialists, the TA "community" can sometimes seem a little introspective, self-regarding and indeed perhaps almost presumptuous about its existence, activities and importance. A good antidote to any such tendencies is for TA practitioners to ask, among contacts in the world outside TA, what these contacts understand is meant by "Technology Assessment". It

DOI: 10.1057/9781137561725.0022

TABLE 11.1 *2nd PACITA Conference programme*

	Fact sheet	1st European TA conference Prague 13 -15 March 2013	2nd European TA conference Berlin, 25 - 27 February 2015
	Date	13–15 March 2013	25–27 February 2015
	Place	National Technical Library, Prague, the Czech Republic	Umweltforum Auferstehungskirche, Berlin, Germany
Participants	Participants	245	349
	Speakers	155	230
	Countries	31	33
	5 Most Represented European countries	Germany – 59 The Czech Rep. – 53 The Netherlands – 26 Austria – 14 Belgium – 10	Germany – 150 Austria – 22 The Netherlands – 21 United Kingdom – 20 Denmark – 15
	5 Most Represented Non-European countries	Japan – 7 Australia – 4 Rep. of Korea – 4 USA – 3 Turkey – 2	Japan – 8 USA – 5 Russia – 3 China – 3 Australia – 3
Sessions	Sessions:	22	42
	Keynote speakers	Wiebe Bijker Stefan Böschen Rut Bízková	Naomi Oreskes Roger Pielke, Jr
	The most discussed topics (As per sessions)	Governance and Participation Technology Assessment Methods Evidence-Based Policy Making Emerging Technologies Ageing and Health Care Big Data and Privacy Sustainable Development Robotics and Synthetic Biology	Responsible Research and Innovation Technology Assessment Methods Governance and Participation Evidence-Based Policy Making Robotics and Synthetic Biology Ageing and Health Care Big Data and Privacy Energy
	Special formats	Panel Discussion/Round Table Politicians' and Researchers' Views on Joint Projects TA Meets Young Talents Author Meets Critics	PACITA Workshop Panel Discussion/Round Table Film Presentation World Café Seminar

Continued

DOI: 10.1057/9781137561725.0022

TABLE 11.1   *Continued*

Outcomes	Web page	pacita.strast.cz/en/conference	berlinconference. pacitaproject.eu
	Social media	Twitter@PACITAproject #paciTA13 Facebook, YouTube	Twitter @PACITAproject #paciTA15 Facebook, YouTube
	Outcomes	Book of Abstracts Conference Proceedings	Book of Abstracts Conference Proceedings

invariably becomes clear that we operate in a rather restricted space, whose recognition by wider society is limited. TA is immanently in a supplicatory relationship with wider society. It has legitimacy, indeed an existential claim, *only if it is seen as having utility by that wider society.*' (Cope, 2014: 376).

Notes

1   All agendas and conference topics can be downloaded here: http://www. openta.net/nta-tagungen (in German).
2   See also http://eptanetwork.org/about.php.

DOI: 10.1057/9781137561725.0022

12

E-Infrastructure for Technology Assessment

M. Nentwich

Abstract: *Nentwich gives an in-depth account of developments within the TA community towards a common e-infrastructure for technology assessment (TA). The author argues that while technology development is genuinely international, there are too few endeavours to address technology assessment (TA) issues internationally; likewise, there are no sustainable online platforms for knowledge sharing, dissemination and public debate as yet. The PACITA project partners therefore worked to establish such an infrastructure by means which the article details. Creating and sustaining a strong, interactive e-infrastructure for cross-European TA is both greatly challenging and worthwhile as it would ultimately help to nuance and possibly even democratize European science, technology and innovation policy. Nentwich therefore argues for the continuation of these efforts by central actors in and supporters of TA.*

Klüver, Lars, Rasmus Øjvind Nielsen, and Marie Louise Jørgensen, eds. *Policy-Oriented Technology Assessment Across Europe: Expanding Capacities.* Basingstoke: Palgrave Macmillan, 2016. DOI: 10.1057/9781137561725.0023.

> *While technology development is genuinely international, there are only few endeavours to address technology assessment (TA) issues internationally; likewise, there are no sustainable online platforms for knowledge sharing, dissemination and public debate as yet. Creating and sustaining a strong, interactive e-infrastructure for cross-European TA is both greatly challenging and worthwhile as it would ultimately help to nuance and possibly even democratize European science, technology and innovation policy.*

Recently, the international TA community started facing this challenge and increasingly produces digital infrastructures for daily work and communication as well as for outreach. This chapter presents elements of current e-infrastructures and practices. A particular focus is on the new TA Portal launched by the PACITA consortium in 2012. This portal has the potential to become a one-stop service and exchange platform for both TA practitioners and those interested in technology policy and TA in general. However, in order to reach and sustain its full potential, this core e-infrastructure for TA needs to become more than a database with interesting and potentially useful content. The article argues that the portal should turn into a dynamic and interactive platform.

We distinguish the following main elements of TA e-infrastructures as they exist today: the EPTA website and project database; videoconferencing tools as used in international projects; outreach activities of TA on social network sites such as Facebook and others; a few TA-related tools and databases; the Network for Technology Assessment's web portal openTA; and the PACITA TA Portal. The core of the latter is a database that covers TA publications, projects, experts, and organizations. Furthermore, the Portal recommends selected TA-related Internet resources and offers a list of the latest TA news on the homepage. The TA Portal is a work in progress; plans to enhance its functionality, described in the following, are being implemented.

By devising the TA Portal, by coordinating the joint international effort to filling the database, and by reflecting the usability and usefulness for future activities, we learned that it is both an enormous challenge in technical, conceptual, and organizational terms, and it is a promising opportunity. While putting in place a schema and (semi-)automatic procedure to fill a database with useful information was (and is) a big effort, it still is only half the story. Turning the Portal into a lively platform that serves

DOI: 10.1057/9781137561725.0023

the TA community and that connects it to its addresses and interested actors across Europe demands a far greater effort. Such a platform would be not only a technical tool but also a social enterprise. In order to activate its content, editing staff is needed with a mandate not only to disseminate results but also to advocate the balanced results reached by TA methods for incorporation into the European debate.

Reaching the full potential of the TA e-infrastructure in the making and scaling it up needs:

▸ An electronic infrastructure for TA practitioners that can also serve as a platform for debate and policy support demands financial resources and time to incorporate lessons learned on a continuous basis.

▸ A permanent cross-European TA network with a sustainable budget to support editorial or facilitating functions.

Introduction

Technology development and diffusion has no borders, nor have impacts, chances, and risks of new technologies. Despite this obvious fact, there are only a few endeavours to address technology assessment issues at the international level (in particular in a series of common EU projects,[1] such as PACITA), but most TA takes place in the national arena. The main reason for this is that technology governance, so far, is to a large extent national; furthermore, assessment is culturally bound and also dependent on local circumstances. Nonetheless, TA practice is increasingly international in the sense that it relies on a network that provides for the exchange of methods and personnel, as well as for mutual stimulation and enrichment when it comes to watching and assessing technology trends. The backbone of this network consists of regular conferences (EPTA, PACITA, NTA, and ITA series), journals, and two associations (EPTA and NTA). In line with, but following with some delay, the global trend towards cyber-science (Nentwich, 2003) and open science (e.g. Bartling and Friesike, 2013), the international TA community increasingly uses digital infrastructures for daily work and communication.

The earliest elements of this evolving e-infrastructure for technology assessment date from the late 1980s and 1990s (cf. Nentwich and Riehm, 2012; Nentwich, 2010). Most prominently, the German 'TA-Databank',

DOI: 10.1057/9781137561725.0023

operated by the ITAS in Karlsruhe from 1987 to 1998 (Berg and Bücker-Gärtner, 1988), was an encompassing online database (still available on CD-ROM). By 1999 it contained datasets of over 570 institutions, approximately 3.400 projects and 7.000 publications.[2] From 1997 to 2013 the ITA in Vienna took care of the virtual library 'TA in the WWW', containing some 270 links.[3] A first attempt to establish a social network for TA practitioners on the basis of the Ning platform in 2008 by the NBT in Oslo attracted only a small proportion of the community (approximately 75 members in 2010; cf. Nentwich, 2010) and never showed much activity (it has been offline since 2013). Furthermore, the German TA network experimented from 2006 to 2012 on its previous website with a meta-search engine (on the basis of Google Custom Search) covering the content of the NTA member organizations' websites. In addition, some EU-funded projects resulted in web platforms offering specific TA- and foresight-related tools and databases (listed in the section below). In the meantime, in particular in the framework of the PACITA project and the NTA network, new developments are under way.

The remainder of this chapter gives an overview of how digital means, mainly via the Internet, are used and needed both inside the TA community and vis-à-vis its addressees in politics and in society today. In the next section, the elements of this infrastructure are briefly described, followed by a longer section on the international TA Portal designed and implemented by the PACITA project team and by a concluding section with an outlook on the development of the e-infrastructure for TA. We argue that an increased online presence of the cross-European TA community would benefit European policy making.

The main elements of the current TA e-infrastructure

From around 2010, actors in the TA community have started new initiatives to build up a modern digital infrastructure. The main fora of these activities are the German TA network (NTA),[4] the European Parliamentary TA network (EPTA),[5] and the EU-funded project Parliaments and Citizens in TA (PACITA).[6] In 2014 the e-infrastructure of the TA community included the following elements:

EPTA website and project database: For more than ten years the website of EPTA features an online project database, now containing almost 900 datasets with titles, keywords, project life spans, contact persons,

DOI: 10.1057/9781137561725.0023

descriptions, and links to further information.[7] The content of the database is provided by the member institutions by more or less regularly filling an online form; the site and database is currently operated by the DBT in Copenhagen – in the future by ITA in Vienna, after a re-launch scheduled for 2015.

Videoconferencing: TA projects are often carried out by dispersed teams with staff from several organizations across Europe. Although TA practitioners also use face-to-face meetings, they have followed the general trend of international professionals by increasingly using videoconferencing tools, such as WebEx (e.g. in PACITA) and most frequently Skype, to meet. While these meetings are considered indispensable for specific purposes or occasions and best practices have evolved over time, experiences with network stability and technical quality of the services are still mixed.

TA on social network sites: As TA has an important interface with the general public alongside the political and the academic spheres, all TA organizations have public websites that communicate their identities and work. Many but not all TA organizations are now also present on the main social network sites, such as Facebook and Twitter. Many also contribute to TA-related topics on Wikipedia (Nentwich, 2010). For most organizations, however, this work takes place with limited success and resources. EPTA and NTA as well as some TA projects like PACITA are also operating Facebook pages. Except for some individuals, Twitter is still used only sparingly by TA organizations or practitioners (cf. König, 2015).

TA-related tools: A few EU-funded projects resulted in databases of platforms serving specific purposes of the TA community. One such example is Doing Foresight,[8] a support instrument for activities/ projects on future-oriented policy analysis. Another is the Decision support on security investment (DESSI) Tool,[9] giving insight into the pros and cons of specific security investments. A third is the European Foresight Platform (EFP), providing briefs of foresight processes carried out in Europe.[10] The main problem with these tools and databases is, that after the end of project-related funding, they tend to be forgotten and not updated anymore. Furthermore, the international publications' repository, in particular the one for economic research papers (RePEc), provide the opportunity to organize TA resources on the Internet (cf. Moniz, 2015).[11]

NTA Fachportal openTA: In the framework of NTA, funded by the German research fund DFG and carried out by ITAS and

DOI: 10.1057/9781137561725.0023

ITAS' partners, the openTA portal is the latest newcomer of the e-infrastructure of TA, which launched in 2014.[12] The main elements of openTA currently are: an NTA members' (individual and organizational) database; a news aggregator, fed by the NTA member organizations; a common calendar of TA-related events (conferences, calls, teaching, lectures, etc.), also fuelled by NTA members; a TA blog; and an encompassing TA publication database that covers publications not only of the member organizations but well beyond the TA community, which is also fuelled by the German national library and other databases. The openTA portal is not intended to be a technology-oriented database project, but rather an 'innovation project for the TA community' (Nentwich and Riehm, 2012, Riehm and Nentwich, 2014).

PACITA TA Portal: Since 2011 one of the tasks of the EU-funded project PACITA was the establishment of a comprehensive portal for TA-related information in Europe and beyond. The task leader was ITA in Vienna. On 22 October 2012, the first version of the new service had been launched at the EPTA Council meeting in Barcelona.[13] The portal cooperates with the openTA initiative with a view to avoid duplication and exploit synergies.

The PACITA TA Portal

The core of this web platform is a database that covers four types of TA-related information: publications, projects, experts, and organizations. The users interact with the database via either simple or more detailed search forms. The results are presented in tabs and as a hypertext, allowing for browsing in the lists of results – for example, by jumping from a publication to its authors or from there to their home organization or to the related project. The users may also directly retrieve a list of the latest updates of the database (recent publications and more). See the following screenshot for an impression of the look and feel of the website.

The datasets are provided in a decentralized way by the participating TA organizations, harvested and stored centrally by the portal. Some of the data providers use automated scripts to transform the content of their local databases into the format prescribed by the portal; others do it manually.

DOI: 10.1057/9781137561725.0023

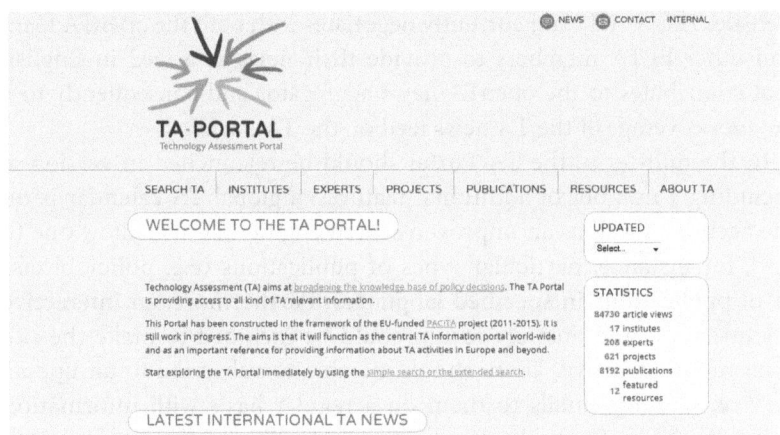

FIGURE 12.1   *Homepage of the TA Portal (screenshot taken on 30 April 2015)*

At the time of writing, the database includes datasets from 17 organizations, over 200 experts, 621 projects, and roughly 8200 publications. While the portal includes data from all PACITA member organizations and two other TA units (the US GAO and the German TAB), it is intended to have global reach, including relevant information from any organization that works in the field of technology assessment. As an obvious next step, further EPTA members (some are already part of the PACITA project and hence of the Portal) shall be included. Furthermore, a (two-way) bridge between the openTA and PACITA portals should be established to include data from further NTA members (some are already part of the PACITA project and hence of the Portal). Aiming to attract more content providers, PACITA has adopted a policy document that sets out in a transparent way the criteria for membership in the Portal. These include a definition of TA and of eligible TA organizations (individual persons cannot directly contribute content to the Portal).[14]

Beyond these core functionalities, the TA Portal has two further features: First, it recommends a few special Internet resources (currently ten, including the PACITA VolTA magazine and PACITA deliverable 2.2 on the comparison of existing PTA organizations). Second, on the homepage, a list of the latest TA news is presented. This is the first outcome of the cooperation between the TA Portal and openTA, as the latter provides a so-called widget to include the aggregated news on any

DOI: 10.1057/9781137561725.0023

website. The portal team currently negotiates with both the openTA team and other EPTA members to provide their news as a feed in English that contributes to the openTA news aggregator and consequently to a broader coverage of the TA news feed on the TA Portal.

In the mid-term, the TA Portal should be relaunched in version 2, including a number of additional features: a global TA calendar is on the agenda as well as an improved search engine that will allow one to find, for instance, particular types of publications (e.g. policy briefs) or of publications in specified languages. Furthermore, an interactive TA questions and answers forum could be included to make the site even more attractive. Users should be able to subscribe to an update service, sending emails to them on a regular basis with information about the latest TA publications or projects. Finally, there is a plan to set up (and include in the search) an open access TA repository for TA-related publications that are not included in one of the member organizations' websites. This would enable researchers affiliated with non-TA organizations, but publishing relevant articles, to include them in the TA Portal.

The way ahead

There is no doubt that broadening the knowledge-base of political decision making is urgent due to the complexity of the grand challenges that our societies face. As argued in the introduction to this volume, TA in its various forms, from providing well-balanced expertise to involving stakeholders and citizens, contributes in effective and well-established ways to future-oriented policy activities. Given the intrinsic cross-border nature of technology development, the need for a strong cross-European foundation of TA is evident. To induce dynamic cooperation, open debate, and knowledge sharing on these highly salient issues the TA community and its addressees will greatly benefit from a state-of-the-art e-infrastructure.

Our brief description of the current digital infrastructure available for technology assessment shows that with the PACITA TA Portal (along with the openTA platform) the TA community is about to reach a next level. The current platform has the potential to become a one-stop service for TA, especially if it is developed further both in terms of the types and quality of services offered and the scope of resources included. The

DOI: 10.1057/9781137561725.0023

PACITA TA Portal in particular could serve as the background infrastructure for the EPTA website.

An Internet portal can be regarded as an infrastructure in two ways. First, it is an internal service that is intended to help TA practitioners to do what they have to do: to stay up to date about the TA literature; to know whom to approach for specific expertise; to build on projects done by others; to stay informed about the current activities of fellow TA units; to be aware of TA events; to stay tuned with current trends; and so on. Furthermore, such an infrastructure may potentially offer a communicative space for exchange, be it written (blogs and discussion fora), spoken (videoconferencing), and possibly even social network functions. So far, the current infrastructure focuses on mainly the internal aspect, while there is still a long way to offer an ideal environment for online collaboration.

The second way to look at such a portal is with the eyes of the customers of TA – that is, actors in both the political and the public spheres who are interested in technology policy and assessment. To turn the existent portal into an information platform that presents TA-related information in a format that is attractive to laypersons in general and to decision makers in particular is, however, a much greater challenge. This would mean adding a public relations side to the sober database; it would mean having an editorial team that selects and presenting the latest TA results in a catchy way; and it would mean making the platform interactive and communicative, which possibly includes having a presence on the popular social network sites. All this needs to be thought and structured as a long-term, sustainable enterprise.

Both aims, the internal and the external one, are worthwhile to invest in, be it in terms of ideas, time or, ultimately, financial resources. The latter will have come to an end with the conclusion of the PACITA project in spring 2015, so the future of the TA Portal and hence the backbone of the current international e-infrastructure for TA is in limbo. Keeping the platform alive will be possible for some time on the basis of contributions made in kind by the leading TA organizations. Expanding it, improving it, and turning it into the envisaged one-stop service and communicative platform for TA, however, can be done only with an additional financial effort and a certain element of (cyber-) entrepreneurship. The TA community is called to make its own modern infrastructure a prime concern. And it needs continuous societal support.

DOI: 10.1057/9781137561725.0023

Notes

1 See Chapter 5.
2 Cf. http://www.itas.kit.edu/1999_008.php.
3 In 2014 this link collection is still available via the EPTA website at http://www-97.oeaw.ac.at/cgi-usr/ita1/tawww.pl?site=epta.
4 http://www.openta.net/netzwerk-ta.
5 http://eptanetwork.org.
6 http://www.pacitaproject.eu.
7 http://eptanetwork.org/projects.php.
8 http://www.doingforesight.org.
9 http://securitydecisions.org/decision-support-tool.
10 http://www.foresight-platform.eu/briefs-resources.
11 http://biblio.repec.org/entry/oca.html.
12 http://www.openta.net.
13 http://technology-assessment.info.
14 http://technology-assessment.info/images/TA-Portal-Policy_v260313.pdf.

DOI: 10.1057/9781137561725.0023

Bibliography

M. Almeida (2012) 'Explorative Country Study: Portugal' in Hennen, L., Nierling, L. (eds), *PACITA Deliverable 4.1. Expanding the TA landscape*, http://www.pacitaproject.eu, pp. 221–54.

M. Almeida (2014) *Policy Hearing report. PACITA Deliverable 5.3*, http://www.pacitaproject.eu

D. Barben et al. (2008) 'Anticipatory Governance of Nanotechnology: Foresight, Engagement, and Integration' in Edward J. Hackett et al. (eds), *The Handbook of Science and Technology Studies. Third Edition* (Cambridge, MA: MIT Press), pp. 979–1000.

M. Barland et al. (2012) *Making cross European TA. Deliverable 2.4 of the PACITA project*, http://www.pacitaproject.eu.

M. Barland (2013) *Scenario Workshop Method Description. PACITA Deliverable 6.1*, http://www.pacitaproject.eu.

S. Bartling and S. Friesike (eds) (2013) *Opening Science. The Evolving Guide on How the Web is Changing Research, Collaboration and Scholarly Publishing* (Berlin, Heidelberg and New York: Springer).

U. Beck (1992) *Risk Society. Towards a New Modernity* (London: Sage).

U. Beck, A. Giddens and S. Lash (eds) (1994) *Reflexive Modernization* (Cambridge: Polity Press).

U. Beck and E. Grande (2007) *Cosmopolitan Europe* (Cambridge: Polity Press).

A. Benz (2007) 'Governance in connected arenas – Political science analysis of coordination and control in complex rule systems', in Dorothea Jansen (ed.),

New Forms of Governance in Research Organizations. From Disciplinary Theories towards Interfaces and Integration. (Heidelberg: Springer), pp. 3–22.

I. v. Berg and H. Bücker-Gärtner (1988) *Aufbau einer Datenbank über Institutionen, Projekte und Veröffentlichungen auf dem Gebiet der Technikfolgenabschätzung (TA)* (Karlsruhe: Kernforschungszentrum Karlsruhe).

W. Bijker, T. Hughes and T. Pinch (eds) (1987) *The Social Construction of Technological Systems: New Directions in the Sociology and History of Technology* (Cambridge, MA andLondon: MIT Press).

D. Bütschi (2012) *Knowledge-Based Policy Making. Report of the First Parliamentary TA Debate, Held in Copenhagen on 18 June* (Berne: TA-SWISS).

D. Bütschi (2014) *Strengthening Technology Assessment for Policy-Making Report of the Second Parliamentary TA Debate held in Lisbon on 7–8 April* (Berne: TA-SWISS).

D. Bütschi et al. (2004) 'The practice of TA; Science, interaction and communication' in M. Decker and M. Ladikas (eds), *Bridges between Science, Society and Policy: Technology Assessment – Methods and Impact* (Berlin, Heidelberg and New York: Springer), pp. 13–55.

D. Cope (2014) 'Technology Assessment and Parliament – the Indispensible Link' in T. Michalek et al. (eds), *Technology Assessment and Policy Areas of Great Transitions. Proceedings from the PACITA 2013 Conference in Prague* (Prague: Technology Centre ASCR).

L. Cruz-Castro and L. Sanz-Menéndez (2005) 'Politics and institutions: European parliamentary technology assessment' *Technological Forecasting & Social Change*, LXXII, pp. 429–48.

M. Decker and M. Ladikas (eds) (2004) *Bridges between Science, Society and Policy: Technology Assessment – Methods and Impacts* (Berlin, Heidelberg and New York: Springer).

P. Delvenne (2011) *Science, technologie et innovation sur le chemin de la réflexivité. Enjeux et dynamiques du Technology Assessment parlamentaire* (Louvain-La-Neuve: Academia L'Hartmattan).

P. Delvenne, C. Fallon and S. Brunet (2011) 'Parliamentary Technology Assessment Institutions as Indications of Reflexive Modernization' *Technology in Society*, XXXIII, pp. 36–43.

P. Delvenne, B. Rosskamp and C. Fallon (2012) 'Explorative Region Study: Wallonia, Belgium' in L. Hennen, L. Nierling (eds), *Expanding the TA landscape. PACITA Deliverable 4.1.* http://www.pacitaproject.eu, pp. 255–86.

P. Dröge (2013) 'Talking TA: PACITA Great Transitions Conference' *volTA*, IV.

C. Enzing et al. (2012) *Technology Across Borders. Exploring perspectives for pan-European Parliamentary Technology Assessment* (European Parliament: Brussels).

European Commission (2010) *Digital Agenda for Europe. A European 2020 Initiative*, http://ec.europa.eu/digital-agenda.

European Council (2009) *The Lund Declaration*, http://www.vr.se.

Expert Working Groups on Public Health Genomics (2013) *Future Panel on Public Health Genomics – Expert Working Group Reports* (The Hague: Rathenau Instituut).

Y. Ezrahi (1990) *The Descent of Icarus: Science and the Transformation of Contemporary Democracy* (Cambridge, MA: Harvard University Press).

G. Falkner, W. Peissl and H. Torgersen (1994) 'PTA in Europa: der Vergleich' in G. Falkner, W. Peissl and H. Torgersen (eds), *Technikfolgen-Abschätzung in Europa, Forschungsstelle für Technikbewertung*, Wien, pp. 166–93.

C. Fitzgerald (2014) *Policy Status Overview. PACITA Deliverable 6.3*, http://www.pacitaproject.eu.

J. Ganzevles and R. van Est (2012) *TA practices in Europe. Deliverable 2.2* (European Union, Brussels).

J. Ganzevles, R. van Est and M. Nentwich (2014) 'Embracing Variety: Introducing the Inclusive Modelling of (Parliamentary) Technology Assessment' *Journal of Responsible Innovation*, I, 3, pp. 292–313.

A. Grunwald (2009) 'Technology Assessment: Concepts and Methods' in A, Meijers (ed.), *Philosophy of Technology and Engineering Sciences* (Amsterdam, Boston, Heidelberg, London, New York, Oxford, Paris, San Diego, San Francisco, Singapore, Sydney and Tokyo: Elsevier), pp. 1003–46.

A. Grunwald (2011) 'Responsible Innovation: Bringing together Technology Assessment, Applied Ethics and STS research' *Enterprise and Work Innovation Studies*, VII, pp. 9–31.

N. Gudowski et al. (2014) 'Responsible Research und TA – Innovationen neu gestalten' in *Bericht der 6. Konferenz des Netzwerks TA und der 14. Jahreskonferenz des ITA* (Wien, Vienna: ITA).

L. Hennen et al. (2004) 'Towards a Framework of Assessing the Impact of Technology Assessment' in M. Decker and M. Ladikas (eds),

DOI: 10.1057/9781137561725.0024

*Bridges between Science, Society and Policy: Technology Assessment –
Methods and Impacts* (Berlin, Heidelberg and New York: Springer).

L. Hennen and M. Ladikas (2009) 'Embedding Society in European
Science and Technology Policy Advice' in M. Ladikas (ed.),
*Embedding Society in Science and Technology Policy – European and
Chinese Perspectives* (European Commission, Brussels), pp. 39–64.

L. Hennen and L. Nierling (2012) *Expanding the TA-Landscape. Country
Studies. PACITA Deliverable 4.1*, http://www.pacitaproject.eu.

L. Hennen and L. Nierling (2014a) 'Expanding the TA Landscape.
Barriers and Opportunities for Establishing Technology Assessment
in Seven European Countries' in T. Michalek et al. (eds), *Technology
Assessment and Policy Areas of Great Transitions. Proceedings from
the PACITA 2013 Conference in Prague* (Prague: Technology Centre
ASCR), pp. 67–73.

L. Hennen and L. Nierling (2014b) 'A Next Wave of Technology
Assessment? Barriers and Opportunities for Establishing TA in Seven
European Countries' *Science and Public Policy*, published first online.

L. Hennen and L. Nierling (eds) (2015) 'TA as an Institutionalized
Practice – Recent National Developments and Challenges' in
Technikfolgenabschätzung in Theorie und Praxis, XXIV, pp. 1.

S. Jasanoff (2005) *Designs on Nature. Science and Democracy in Europe and
the United States* (Princeton, NJ: Princeton University Press).

S. Joss and S. Bellucci (eds) (2002) *Participatory Technology
Assessment – European Perspectives* (London: Centre for the Study of
Democracy (CSD) at University of Westminster in association with
TA Swiss).

L. Klüver et al. (2004) 'Technology Assessment in Europe: Conclusions
and Wider perspectives' in M. Decker and M. Ladikas (eds), *Bridges
between Science, Society and Policy. Technology Assessment – Methods
and Impact* (Berlin, Heidelberg and New York: Springer).

L. Klüver (2014) 'Foreword' in T. Michalek et al. (eds) *Technology
Assessment and Policy Areas of Great Transitions. Proceedings from
the PACITA 2013 Conference in Prague* (Prague: Technology Centre
ASCR).

R. König (2015) 'Tweeting TA – Chances and Pitfalls of Microblogging
in Technology Assessment' Paper at the 2nd European TA
Conference: The Next Horizon of Technology Assessment, Berlin, 25
February.

J. Kooiman (2003) *Governing as Governance* (London: Sage Publications).

DOI: 10.1057/9781137561725.0024

V. Kozarev (2012) 'Explorative Country Study: Bulgaria' in L. Hennen and L. Nierling (eds), *Expanding the TA landscape. Deliverable 4.1.* http://www.pacitaproject.eu, pp. 30–62.

A. Krom and D. Stemerding (2014) *Future Panel Method Description. PACITA Deliverable 5.4*, http://www.pacitaproject.eu.

B. Latour (1993) *We Have Never Been Modern* (Cambridge, MA: Harvard University Press).

B. Latour and S. Woolgar (1979) *Laboratory Life: the Construction of Scientific Facts* (Princeton, NJ: Princeton University Press).

E. Leichteris and G. Stumbryte (2012) 'Explorative Country Study: Lithuania' in L. Hennen and L. Nierling (eds), *Expanding the TA landscape. PACITA Deliverable 4.1*, http://www.pacitaproject.eu, pp. 184–220.

R. Mayntz and F. Scharpf (1995) 'Steuerung und Selbstorganisation in staatsnahen Sektoren' in R. Mayntz and F. Scharpf (eds), *Gesellschaftliche Selbstregelung und politische Steuerung* (Frankfurt am/ Main and New York: Campus), pp. 9–38

U. Meidert and H. Becker (2013) *Telecare technology in Europe. PACITA deliverable 6.2*, http://www.pacitaproject.eu.

T. Michalek et al. (2014) 'Introduction' in T. Michalek et al. (eds) *Technology Assessment and Policy Areas of Great Transitions. Proceedings from the PACITA 2013 Conference in Prague* (Prague: Technology Centre ASCR).

A. B. Moniz and A. Grunwald (2009) 'Recent Experiences and Emerging Cooperation Schemes on TA and Education. an Insight into Cases in Portugal and Germany' *Technikfolgenabschätzung – Theorie und Praxis*, XVIII, pp. 17–24.

A. Moniz (2015) 'The Use of International Repositories and the Definition of the TA Research Field: The Case of RePEc' Paper at the 2nd European TA Conference: The Next Horizon of Technology Assessment, Berlin, 25 February 2015.

J. Mosoni-Fried, A. Zsigmond and E. Palinko (2012) 'Explorative Country Study: Hungary' in L. Hennen and L. Nierling (eds), *Expanding the TA landscape. PACITA Deliverable 4.1*, http://www.pacitaproject.eu, pp. 104–43.

M. Nedeva and M. Stampfer (2012) 'From "Science in Europe" to "European Science"' *Science*, CCCXXXIII.

M. Nentwich (2003) *Cyberscience: Research in the Age of the Internet* (Vienna: Austrian Academy of Sciences Press).

DOI: 10.1057/9781137561725.0024

M. Nentwich (2010) 'Technikfolgenabschätzung 2.0' *Technikfolgenabschätzung – Theorie und Praxis*, XIX, 2, pp. 74–79.

M. Nentwich and U. Riehm (2012) 'Internationale Fachportale für Technikfolgenabschätzung – Brauchen wir eines oder sogar mehrere?' *Technikfolgenabschätzung – Theorie und Praxis*, XXI, 3, pp. 76–79.

L. Nierling et al. (2013) 'The International TA Community Comes Together – Once Again Please!' *TATuP – Technology Assessment, Theory and Practice*, XXII, 2, pp. 101–05.

E. D. H. Olsen and H.-J. Trenz (2010) 'Deliberative Polling. a Cure to the Democratic Deficit?' Arena Working Paper, http://arena.uio.no.

P. O'Reilly and F. Adam (2012) 'Explorative Country Study: Ireland' in L. Hennen and L. Nierling (eds) *Expanding the TA landscape. Deliverable 4.1*, http://www.pacitaproject.eu, pp. 144–83.

R. Owen, J. Bessant and M. Heintz (eds) (2013) *Responsible Innovation* (London: John Wiley).

PACITA (2014) *Strengthening Technology Assessment for Policy-Making, 7–8 April 2014 in Lisbon in the Portuguese Parliament*, http://www.pacitaproject.eu/wp-content/uploads/2014.

O. Pokorny, L. Hebakova and T. Michalek (2012) 'Explorative Country Study: Czech Republic' in L. Hennen and L. Nierling (eds), *Expanding the TA Landscape. Deliverable 4.1*. http://www.pacitaproject.eu, pp. 63–103.

M. Rask, R. Worthington and R. Lammi (2012) *Citizen Participation in Global Environmental Governance* (London and New York: Routledge).

D. Riedlinger (2013) *TA Meeting in Prague*, http://www.oeaw.ac.at/ita/en/events/event-news/voller-erfolg-in-prag, 20 March.

U. Riehm and M. Nentwich (2014) 'Was Sie schon immer TA wissen wollten: Zwei neue Webportale' *TAB-Brief*, XLIV, pp. 49–50.

N. Slocum (2003) *Participatory Methods Toolkit* (Brussels: the King Baudouin Foundation and the Flemish Institute for Science and Technology Assessment (viWTA) in collaboration with the United Nations University – Comparative Regional Integration Studies (UNU/CRIS)).

D. Stemerding and A. Krom (eds) (2013) *Expert Paper for the Future Panel on Public Health Genomics. PACITA Deliverable 5.1* (The Hague: Rathenau Instituut).

D. Stemerding and A. Krom (2014) *Policy Brief on Public Health Genomics. PACITA Deliverable 5.2* (The Hague: Rathenau Instituut).

DOI: 10.1057/9781137561725.0024

TNS Opinion & Social (2005) *Special Eurobarometer 224/ Wave 63.1. Europeans, Science and Technology* (Brussels: European Commission).

TNS Opinion & Social (2010) *Special Eurobarometer 340/ Wave 73.1. Science and Technology* (Brussels: European Commission).

TNS Opinion & Social (2013) *Special Eurobarometer 401. Responsible Research and Innovation (RRI), Science and Technology* (Brussels: European Commission).

J. Van Eijndhoven (1997) 'Technology Assessment: Product or Process?' *Technological Forecasting and Social Change*, LIV, pp. 269–86.

R. Van Est and F. Brom (2012) 'Technology Assessment: Analytic and Democratic Practice' in R. Chadwick (ed.), *Encyclopedia of Applied Ethics: 2nd ed. Vol. 4* (San Diego, CA: Academic Press), pp. 306–20.

N. J. Vig (2000) 'The European Parliamentary Technology Assessment Experience' in N. J. Vig and H. Paschen (eds), *Parliaments and Technology. The development of Technology Assessment in Europe* (New York: State University of New York Press).

N. J. Vig and H. Paschen (2000) *Parliaments and Technology. The Development of Technology Assessment in Europe* (New York: State University of New York Press).

R. von Schomberg (2012) 'Prospects for Technology Assessment in a framework of Responsible Research and Innovation' in M. Dusseldorp and R. Beecroft (eds), *Technikfolgen abschätzen lehren. Bildungspotenziale transdisziplinärer Methoden* (Wiesbaden: Springer).

R. von Schomberg (2013) 'A Vision of Responsible Innovation' in R. Owen, J. Bessant and M. Heintz (eds), *Responsible Innovation: Managing the Responsible Emergence of Science and Innovation in Society* (Chichester, UK: John Wiley & Sons).

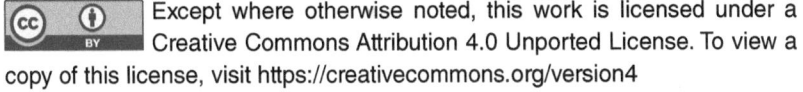
DOI: 10.1057/9781137561725.0024

OPEN

Index

DOI: 10.1057/9781137561725.0025

DOI: 10.1057/9781137561725.0025